Boltzmann's Tomb

Boltzmann's Tomb

TRAVELS IN SEARCH OF SCIENCE

Bill Green

BELLEVUE LITERARY PRESS

NEW YORK

First published in the United States in 2011 by
Bellevue Literary Press, New York

FOR INFORMATION ADDRESS:
Bellevue Literary Press
NYU School of Medicine
550 First Avenue
OBV 640
New York, NY 10016

This book was published with the generous support of Bellevue
Literary Press's founding donor the Arnold Simon Family Trust,
the Bernard & Irene Schwartz Foundation, and the
Lucius N. Littauer Foundation.

Cataloging-in-Publication Data is available from the Library of Congress

Book design and type formatting by Bernard Schleifer
Manufactured in the United States of America
FIRST EDITION
10 9 8 7 6 5 4 3 2 1
ISBN 978-1-934137-35-2 HC

For Wander

Contents

List of Illustrations

A man's work is nothing but the slow trek to rediscover through the detours of art those two or three great and simple images in whose presence his heart first opened.

—*Albert Camus*

Randomness rules our lives.

—*Leonard Mlodinow*

McMurdo Station
and the
Dry Valleys

It was my ninth trip to McMurdo Station, the largest United States base in Antarctica. The base crowded down to the edge of the Ross Sea, and from shore you could see the white mountains of Victoria Land rising like a vision of heaven. I liked the small group I had chosen for this trip and thought we would be perfect for the two lakes we had come to study. Joe was an older chemist, a little overweight, perhaps, and not in the best of health. Through several weeks, I had had to argue with the National Science Foundation that he would be able to do the field studies I had mapped out for him. I had never seen anyone so enthusiastic about an expedition as Joe. At Los Angeles Airport, wearing a cap that counted down the minutes and seconds until the new millennium began, I think he told everyone at the bar and on line for the long flight to Auckland that he was going to Antarctica. If his listener showed even a hint of interest, he would try to explain our entire project. The closer we got to takeoff, the more a smile of expectation lit his face.

My student, Joseph, was not nearly as expressive, but he had worked with me for several years in the lab and was immersed in the research. He was fastidious and had a cheerful alliance with small motors, batteries, copper wires, and instruments of all kinds that would make him invaluable in the unpredictable conditions of the Dry Valleys.

Experience had taught me that nothing ever went according to plan—pumps and generators broke down, drill bits snapped and the one thing you had not thought to bring in duplicate fell into the deep waters of the lake and would never be seen again. But one thing above all else was true: you could always count on Joseph's ingenuity and his data.

Katy, my daughter, who had wanted to go on one of these journeys since she was barely old enough to miss my long absences, was the fourth member of our group. On college breaks, she had lived in huts along the shore of Brazil's Salvador, had contracted malaria in the streets of Accra, and had taken on the winter snows of Colorado and Wyoming in her small mountain tent. She was a student in my geochemistry course and so had a broad overview of that science. I had no doubt she, at twenty-one, was ready for this.

I did, however, have a few questions about myself. Two years earlier, I had had bypass surgery following a heart attack. The surgeon, who was tall and confident, a former navy commander, said, "We'll have you in better shape than you've been in twenty years." I wanted to believe it was true; today, I felt strong and ready for our journey south. The world in its vastness and beauty seemed open to me once again.

&

The physical work seemed almost routine by now: dragging the sleds over the rough, wind-blasted ice on Lake Hoare; setting up the peristaltic pump and the Tygon tube so we could collect samples every few centimeters down through the water column; measuring the dissolved oxygen, the conductivity, and the pH. At Lake Fryxell, the hole over the deepest water had already been drilled by another research team so the big rig with its black, screwlike blades and the fifteen feet of extensions that we usually needed for the thick permanent ice was not necessary.

Over the years, the work had become easier. There were several groups in the Taylor Valley now, and there were a few semipermanent structures that had been set up for lab work and for cooking hot meals.

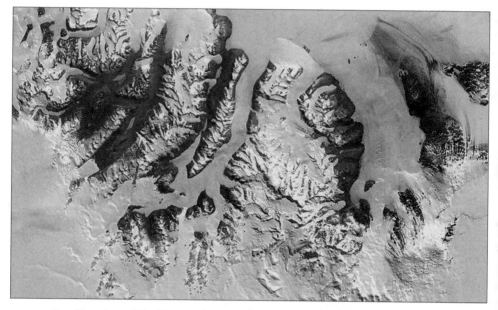

Satellite view of the McMurdo Dry Valleys. Ice-free areas are dark.

As I looked out over the valley from the glistening lake ice, I remembered the emptiness of earlier days, when I could imagine there had never been anyone else here, that we were the first explorers, and the land had welcomed us in its silence and majesty, its bare stone mountains rising into cloud. There was, however, a kind of scientific satisfaction in seeing the growing interest in the lakes, that they were no longer the mere biological and geochemical curiosities they had once been. In a kind of paradigmatic transformation, they had become viewed as mosaics of interconnected glaciers, streams, soils, lakes, microorganisms, and sediments, accessible and tractable model systems, the only lakes in the cold deserts of the earth, and the only models of what the waters of Mars might have been like. They were extreme environments and they hummed with a new importance that extended even beyond this earth.

The sample collecting was always slow, and you had to be obsessed with every step to prevent contamination. In addition to the nutrients nitrate and phosphate, we were studying a group of metals near the middle of the periodic table, metals that are rare in natural waters. Among them were cobalt, copper, nickel, and zinc. These are essential micronutrients, but at higher concentrations they can be lethal. Even the sparse biota of these lakes, which was comprised mostly of algae and bacteria, depended on the metals being in perfect balance. I was interested in what maintained that balance. Over in Lake Vanda, just across the mountains in the Wright Valley, we thought we had seen good evidence for the role that tiny particles of iron oxides and especially manganese oxides played in removing potentially toxic trace substances. Oxides were brought each year by the flowing streams that came in springtime from the glaciers. But who knew what we would find in Lake Fryxell and Lake Hoare; who knew what we would find next year in Miers and Joyce? We proposed that the same processes we had found in Vanda would be operating in all the lakes, and, perhaps, in natural waters everywhere. But this was a long, long extrapolation. Lakes were as different, in fine detail, as the waves of the sea.

After a twelve-hour day on the surface ice, I would be ready for a hot meal and some time in my tent, where I could update my field notes and write in my journal. Katy and Joseph would often go climbing in the Asgards and photograph the valley all the way down to Lake Chad from high in the mountains. Joe, unfortunately, had had irregular heartbeats early in the season and was sent back to Christchurch for examination and then on to Ohio. The National Science Foundation's representatives at McMurdo Station were very cautious and very strict about the regulations they set, whether for health or conduct. If you breached them in any way, you were usually on a plane headed north.

It took us a week to sample the water column in the inch-by-inch way we had planned. When we were through, we called for a helicopter, loaded our water samples onboard, and headed for McMurdo. Katy asked the pilot if he would mind flying near the shrinking ice edge of the Ross Sea, and he agreed. So our return took us along the retreating margin where ocean laps the remaining ice of spring. We flew close to the water so you could see leaping killer whales, tiny Adélie penguins whose lives seemed to be always in peril, and leopard seals sunning themselves in the cold radiance of November sunshine. Katy clapped her hands at this, delighted by the whole impossible dance of beauty and death that revealed itself below. At this time of year, all along the vast stretches of maritime Antarctica, where the simple conversion of ice to water yields the prospect of food, there is imminent danger.

When we arrived back at McMurdo and were unloading the chopper, a woman came up to me and asked if I was Bill Green. I said yes. She asked if I would sign a copy of *Water, Ice and Stone,* a book I had written several years before, and explained that she had read it before coming here. I wrote an inscription on the title page and handed it back, smiling, and told her how pleased I was that she had enjoyed it and had found inspiration for her own journey.

Later, when we were having a hot meal and enjoying the warmth and genial buzz of conversation in the dining hall, Katy asked me a curious question. It might have been related to the unexpected book signing down at the helicopter pad, but in the dining hall it seemed to come out of nowhere. "Why did you decide to work down here, Dad?" she asked, casually sprinkling some salt on a plate of steaming vegetables. She was looking down at her plate, and I probably could have gotten off with a shrug or that palms-open gesture that nicely substitutes for an answer you might not want to give. But I told her simply that I loved the place and I loved the science and had loved it from the very first time I came here. I said, "Most of the people you see around here and over at the Crary Lab and in the metal- and woodworking and electronic shops probably feel the same way I do and want to come back as often as possible. The place gets in your blood. I know you found that in Africa. It's 'Antarctic fever,' and it's powerful, and you never want to think of the day when it will all end. But that day will come." It was not a good answer; even in my book I had not answered that question, although I think I gave ample hints. She seemed to accept this, though, and nodded thoughtfully. Joseph, too, seemed to take this as a reasonable response.

❧

The world mixes and mixes, and pleasures in its task. The unmixings are few. In the midst of the Pacific there is no tasting the waters of the river you love. They are all one now, joined in a mighty watery union whose vastness is the churning of time, past and present, and of all that comes from every land and from the air's infusions and what is far below, where even the fiery rocks churn and make the sea what it is. Oceanographers sometimes call it the "world ocean" since it is so well mixed.

You can see this in the sky, too, the swirl of gases, the darkness bleeding into light. When Mount Pinatubo on Luzon erupted, what came from its mighty throat was not sent into the sky above that island alone, but mixed worldwide across the stratosphere—gases and particles, the oxides of carbon and sulfur and all the pulverized pure stone that would live for years as a thin drawn curtain against the sun.

With lakes it is not so dramatic. They are more humble in scale. Even Lake Superior would be little more than a drop in the world oceans and hardly more than that on the East Antarctic Ice Sheet. But in the landscapes in which they are settled, lakes are the great integrators, the mixing bowls to which every trace of land and water in their domain is brought not to be organized but to be coalesced and commingled as if by some loving chef, brought into a grand synthetic union that is unique among the waters of the world. Acton Lake, near our home in Ohio, is set in the midst of corn and bean fields, and though it is used for sailing and swimming, its waters are defined by the highly fertilized watershed in which it sits. So there is always concern about the phosphates and nitrates that run freely from the fields during spring planting and the havoc these essential nutrients could cause in expanding—exploding, really—the algal population of the lake. Lake Vanda over in Wright Valley, on the other hand, is in a rocky landscape of moraines and sparse—though far from lifeless—soils and has never had, anywhere in its watershed, a single permanent inhabitant nor grown a single ear of corn. Analyze the Onyx River and you can scarcely detect the phosphates and nitrates being brought to the lake. Since Vanda's biota is thin and undernourished, its waters are among the clearest on earth. After a long day of sampling, who can resist a cold cup of Vanda water, unfiltered and direct from the source?

Every lake, like every pond, puddle, and ocean, has admixed with its surface the gases of the atmosphere. We tend not to think of this because it is subtle and slight, but in a way the earth's atmosphere is part of the lake's watershed, though a part that dwells above it and whose constituents are not delivered by any stream. There is a law that governs how much gas will enter—be dissolved in—the surface waters, and we have known it for many years. It was discovered by William Henry of Manchester and presented as an article in *Philosophical Transactions* in 1803. One of Henry's more important findings was that the amount of a particular gas absorbed by a liquid depends on the partial pressure of that gas above the liquid. So regardless of what else might be going on in their watersheds, the high-elevation lakes of Rocky Mountain National Park have lower dissolved-oxygen concentrations than Acton Lake, whose elevation is a mere seven hundred feet. There are, of course, other factors involved in gas solubility: the nature of the liquid, whether benzene or water or some exotic molten salt, and the temperature. Water can usually accommodate more of a specific gas at lower temperatures. The infamous chlorofluorocarbons, which have small, though hardly insignificant, partial pressures in the lower atmosphere, are present at detectable concentrations in the waters of Lakes Fryxell and Hoare. However, like most of the gases in the Dry Valley lakes, they are present at higher levels than those predicted by Henry's law. A kind of extrusion process forces gas out of the ice and into the water as the bottom of the ice layer begins to freeze.

※

Several years ago, I had a most unpleasant reminder of these trace gases and their presence in the atmosphere. It was October and I had decided to start our field season early to measure the metal concentrations in Lake Vanda before the Onyx River began to flow from the Lower Wright Glacier. The temperature in Wright Valley was so low that I could not fall asleep even in my parka and wind pants. I stuffed plastic bottles filled with hot water into my pockets; somehow, then, I was able to doze off. The tent gradually warmed with my body heat and I could sleep fitfully through the night.

The lake shore was steep near our sampling site, so we had to set up camp more than two miles to the east. The days were always sunny and there was barely a cloud. Blue skies, day and night, in the great frigid desert of the McMurdo Dry Valleys. In brilliant sunshine, the ice refulgent, everything covered but our faces, we collected whatever samples we could from a hole whose diameter continued to shrink even as we worked. I had never seen this happen in all the years I had been to the lakes: a twelve-foot-long hole slowly closing in on itself as ice accreted on its sides.

We stayed in Wright Valley for only a week. The wind from the Polar Plateau and the bitter cold seemed to be saying, "Come back when the spring is further along." When I got back to the Crary Laboratory at McMurdo someone asked, "What happened to your face?" I said, "I have no idea. It is feeling a little raw, though."

Then I looked at myself for the first time since we had left the base. My forehead and nose glowed red. My lips were swollen and blistered. It was like one of those days when you ski too long in the mountains of the West and the cold deceives you into thinking there is no need for zinc oxide or any other kind of protection against the sun. But here it was much worse. Then I remembered: in October, we were near the minimum in the ozone curve. The concentration of ozone is lowest at this time of year, and the usual protection that the earth's stratosphere offer against ultraviolet radiation is greatly diminished. People call it the ozone hole, but it is really a thinning in the number of ozone molecules in the stratosphere, and this thinning is most acute over Antarctica.

I couldn't help but replay the entire, nearly disastrous story in my mind as I looked at my singed face. When they were first synthesized by Thomas Midgley in 1928, no one expected that chlorofluorocarbons (CFCs) would be anything but a boon to humanity. They replaced ammonia, a poisonous gas, in refrigerators and opened a new way to home refrigeration beyond the simple icebox of my own memory. They can be used in automobile air conditioners and as additives to foam to make it a better insulator. These compounds were believed to be as inert and harmless as the atoms of argon, and nearly all of us have used them. They are in our blood.

But by the 1970s, questions were being asked and experiments run. James Lovelock, aboard the research vessel *Shackleton* on a cruise from London to Antarctica, found that he could detect CFCs everywhere he sampled in both the Northern and Southern hemispheres. Calculations by Sherwood Rowland suggested that nothing in the lower atmosphere was causing these compounds to break down. Here in the troposphere, where we live, these seemingly innocuous gases were immortal. But, Rowland argued, they must be going somewhere. They must mix upward. With this hypothesis, Rowland and his colleague, Mario Molina, conducted a series of now-classic experiments in which they showed that CFCs had a far more interesting, and possibly darker, life in the stratosphere. A life which for us was ominous. As it turned out, the conditions of the stratosphere were just right to induce the breakdown of CFC molecules. A photon of ultraviolet radiation had just the right energy to sever the chemical bond between carbon and chlorine, and in so doing, set free a highly reactive chlorine atom. This in turn could go on to destroy an ozone molecule—destroying thousands and thousands of them in time. One chlorine atom, recycled over and over, was a catalyst and a destroyer of molecular worlds. Ultimately, this was a complex reaction with many steps, whose rapidity was assured by the presence of polar stratospheric clouds high above Antarctica. The tiny particles that comprise these clouds provide surfaces on which the speed of the reaction is greatly enhanced. It was my misfortune to be wandering the valleys just as these reactions above me were in their very prime.

The various CFC molecules had dissolved in the lakes of the Dry Valleys and exceeded their Henry's law equilibrium values in the cold waters beneath the ice. No doubt they were mixing, even now, in either

21

that hop-jump, zigzag way of molecules, diffusing downward from the top layer to the stratified bottom, or, more quickly, by advection from their source. By now they had entered the ancient lightless zones, zones which had formed long before Thomas Midgley was born, before he had ever thought of synthesizing CFCs in his Dayton lab. But as far as my own research was concerned, my own sampling and measurements, CFCs were on the periphery. I just wanted to know about the metals, their fates, what the particles of iron and manganese mixed into lakes from streams and soils each year were doing deep down there in another of the many provinces of geochemistry, hidden far from our sight.

※

The atmosphere was just a thin nothing when you thought about it, a vaporous band like morning dew, and you could change it in a moment without knowing what you were doing, until someone with a meter on the tossed deck of a ship or on some distant mountain in the middle of the sea said, "Let's check," and everyone wondered, "Why bother, who cares?" But gradually, things seemed to be getting strange: the hurricanes stronger, the oceans warmer, the glaciers and ice fields high in the mountains that people had depended on for centuries, smaller. And down here at McMurdo, the reports were not good, either. Over dinner in the field, we often talked about these things, Joseph and Katy and I, after our samples had been collected and the equipment was safely tucked away in the Scott tent we had set up on the lake. Tomorrow we might hear some more news about the Larsen B Ice Shelf, far to the northwest of us. It would not be good.

The news coming from the atmosphere was so different from what we had learned about the lakes, especially about the metals. All of our results seemed to say that lakes can cleanse themselves; nature has provided for this; the metals visit their waters for only a short while and then are lost to the sediments, buried there, their concentrations never becoming threatening or toxic. And we thought we had traced this down to particles and even knew their identity and their origin. Without huge disturbances in the watershed, there was no need to be concerned that cobalt and cadmium and their cousins would ever pose a threat. Their lifetimes were short. This was good news, we thought, for natural waters everywhere. By contrast, the gases we were spilling into the atmosphere would be there for a very long time—for generations. After all, the atmosphere was just a bright blue thread. I had the burns to prove it.

❀

The transition metals—the long rows of elements strung between the left and right columns of the periodic table—were nothing to be trifled with. They had helped build the ages of humanity, not the least of which was our own: long rails and freight cars that joined mine and mill, the girdered buildings rising straight up from city streets, the engines of jets and rockets long gone to Ganymede and Io. We had been studying these metals for years in this hermitage of wonder and cold that stuck to the soul. I am speaking of the metals iron, manganese, cobalt, nickel, copper, and many others, all now nicely arranged according to atomic number, rather than to atomic weight, as Dimitry Mendeleev had first done. In the waters of the streams and lakes, the metals moved with the randomness of a Boltzmann gas. Building on centuries of tests and designs, on thought that had trawled among stars and stone, we wanted to find the order here. Through days that were sometimes fruitless and barren, too cold to work, we could trace our studies far into the past, discern in the shrouds of history Galileo and Kepler, Svante Arrhenius and Antoine Lavoisier, all those pioneers who, in a sense, were here before us.

Even though there were just three of us doing the geochemistry, it took us only a month to finish all the profiles for the metals, nutrients, and so-called major ions, which are the most common dissolved substances in natural waters—calcium and bicarbonate, and so on. Joseph, who had no class schedule, agreed to stay on for a few weeks longer than Katy and myself, in order to close down our lab at the Crary and to prepare the metal samples for shipping. Katy and I flew out together on the same Herc, stayed for a few days in Christchurch, and then in Auckland. We were back in Ohio in time for the new semester.

On the plane, I wrote a few random thoughts in my notebook. Maybe I could use them someday:

There is enough void here to speak only of void. The land is en-circled with nothingness. Like the hearts of atoms. Emptiness. And born from emptiness. Do you believe that chair exists? It comes from stars. Everything comes from stars. In the fierce burning. In the collapse and burning again. Iron in the streams, in the lakes, every other metal—some immortal wind brought them here. Should you think it all void, here, kick this stone or freeze in these thousand blowing snows. Elements dance on the face of Mendeleev's table. The whole physical world is here. I could not see a single star the whole time, just a ghost moon riding paper thin above the ridges Katy and Joseph climbed. The plane from New Zealand, they say, tried to ascend just before the cone of Erebus, but could not do it. The bodies lay scattered: star-carbon, tears, memories. Of what is grief made? What elements combine? Here is the real void, I think, the one you cannot fill. No matter.

❊

One snowy day in Ohio, I walked about a mile from our house past the cabin built by Zachariah and Elizabeth DeWitt, the oldest extant pioneer homestead in the state. It is made of log and stone and is steady as a starlit night, even after two hundred years. I imagine the DeWitts with their nine children cutting ice from the stream that runs nearby under a gray sky. Past the cabin there is a pasture, where the horses stand blanketed and cold. Sometimes I remember to bring them carrots. They come to the fence and look at me with their large aqueous eyes. The trail that I take leads past the stubble of cut corn and scattered snow, and the sky opens up into a canopy of dark cloud.

Next year's work on Lakes Miers and Joyce would be with a different group. I expected, however, that Joseph would want another season before he headed off for the University of Hawaii. In the meantime, there would be data to analyze and papers to write.

West Lafayette
and Duino

My brother, Eugene, used to say the country was shaped like a deep bowl and no matter how hard you tried, you would always end up somewhere near Indianapolis. In my case that was literally true, because this story begins in West Lafayette, sixty miles to the north of that city, and ends in Oxford, Ohio, a hundred miles to the southeast. I wish he could be here to see just how accurate he was.

This is not a book about the great Austrian physicist Ludwig Boltzmann, nor, despite its importance in my life, is it about Antarctica. It is more about time and chance and the images and dreams we bring with us from childhood which shape who we are and what we become. And it is about science and atoms and starry nights and what we think we remember, though we may have made it up. In the telling, it encompasses many decades and many names and places, some familiar and some obscure. In retrospect, I view it as the story of a journey and a quest.

Boltzmann was not unknown to me when I arrived in West Lafayette. For at least a semester in Pittsburgh, when I was immersed in undergraduate chemistry, his name came up from time to time, but, as is often the case in the sciences, it appeared as an eponymous law or method, utterly impersonal: the Stefan-Boltzmann law or the Maxwell-Boltzmann Distribution. The idea that behind the name there was an actual human being never occurred to me. The equations themselves were enough to

27

get me through exams, even though their full import eluded me. I had little curiosity about the man. Until, that is, the morning I heard Walter Edgell's lecture.

As in so many of the days in that September and October, the skies above the flatness of northern Indiana were a perfect blue. Not far from campus, there was a ripening in the fields, the corn well above my head. When I walked only a short distance in from the road, a labyrinth of green quickly disoriented me. The nights were clear and there was a slight chill in the air. Each day, in that first year of graduate school, there was a sense of newness.

Edgell's lecture had little to do with Boltzmann. It concerned, instead, the visionary work of the Danish physicist Niels Bohr and what he had contributed to our understanding of the atom. How he had seen the electron move in its orbit around the single proton of the hydrogen nucleus, much like a tiny Mars or Venus, until, without warning, it jumped, and in jumping sent off a burst of visible energy into the world. In this way, the beautiful spectrum of hydrogen, with its lines of violet, blue, green, and red, was born.

Edgell's approach was to reveal the history of atomism in fifty-minute installments, like some serial thriller, each episode ending with an insoluble theoretical or empirical problem that left us all wondering, "How will they get out of this one?" If it is possible to be held on the edge of your seat in a physical chemistry lecture, Edgell's dramatic telling did the trick.

But on this particular autumn day, who knows why, some quirk of memory caused Edgell to veer off from his practiced course and toss out an aside obviously missing from his notes. In this moment of transition, there seemed to be a change in his voice, a softening, a seriousness, and his expression visibly saddened. In the midst of his elaboration on the Bohr model, he began to speak about Ludwig Boltzmann.

By the time Bohr's hydrogen model was announced, Boltzmann was dead. But in the late nineteenth century, he had been a defender of atomism, a position that was highly unpopular among physicists. Why was it necessary, they argued, to resort to these unseen, ghostly, almost

Headstone at Boltzmann's Tomb, Vienna, Austria.
Note the equation for entropy near the top of the photo.

occult entities to explain the behavior of matter? Wasn't that just complicating things? But Boltzmann, despite fierce resistance from even his closest colleagues, held to his beliefs and, in fact, considered his contributions to the atomic theory the center of his life's work. Whether it was this rejection, which he felt so personally, or his long battle with manic depression that led to his suicide in Duino, Edgell never gave his opinion on. Nor did he reveal to us how Boltzmann had done it. What he did fix in my mind was an image of the tombstone beneath which Boltzmann lay buried. There is a single equation carved in the stone:

$$S = k \log W$$

Edgell reminded us what each term meant: S is the symbol for entropy, the measure of disorder, of molecular randomness, of the world wearing down, descending a chute it cannot ever reascend; k is a constant, known, honorifically, as the Boltzmann constant; and W is the probability of a particular state or assemblage.

When I walked out of the lecture hall, I realized that something had happened. How these ideas were related was not at all obvious to me, but I resolved that morning to learn more about this mysterious equation and its significance. I had just seen in the thin sketch of a Viennese genius what it meant to be a scientist, what it meant to be life-and-death passionate about an idea, a process of thought. I wanted to understand more about this man and the face of science that he represented, a face that was expressive of joy and hope and crushing disappointment. A face that to me, and to most of us who have gotten our science from textbooks, was all but unknown.

And, as much as anything, I wanted to someday visit Boltzmann's Tomb.

❧

It would take me a life's journey to get to Vienna. If you were to trace my path, it would be like one of those gas molecules that Boltzmann described, traveling an instant through space, colliding, taking a new direction, colliding again, endlessly, over and over, its path unpredictable. I spent a year and a half in West Lafayette before I dropped out of graduate school and moved west to work in orange groves and on small ranches east of Los Angeles. But in that short time in Indiana, I came to know what I liked about chemistry and what I didn't. The onerous memorization was of little interest to me. But the glistening of glass on a laboratory bench, reflecting sunlight from the high windows, seemed as lovely as the coruscations of winter ice on the bare branches of a maple. And the inorganic molecules, with their electrons and orbitals, their clear geometries crafted out of air, seemed the purest works of imagination. I came to love the inorganic, the distant and strange and inhuman, like the poet Robinson Jeffers's "lovely rock," the one that was bearing up the whole California mountain above it, that would outlast the poet, outlast his sons, outlast us all. The entanglements of chemistry, astronomy, physics, and geology—the sciences of molecules, stars, and mountains—were what fascinated me in those days.

There were frequent conversations that semester with the students who had been in Edgell's lecture, but who rarely mentioned Boltzmann. Buckley, who was studying organic chemistry, had written more than a hundred poems and wondered aloud why he had not majored in English and gone off somewhere to write. In winter, with his long black coat and his black beard that fell to his chest, he looked like a young Tolstoy or Whitman and perhaps thought he was. Surely someday he would write something inspired by chemistry, in which he would bring the molecular beauty of crude oil or the snap of ring closure reactions to the attention of the waiting masses. At least, I thought, he would become a professor of literature, teach Kafka's *Castle* or the works of Rilke and Wallace Stevens, all of which seemed to excite him more than anything that was happening in science. Instead, he went on to become a highly successful organic chemist, combining this reagent with that until just the right product emerged. Something you could use.

And Harv wrote plays and show tunes and talked rapidly and loved the ideas of physical chemistry and understood Boltzmann's theories far better than I did. But he hated, or maybe feared, the laboratory, and sometimes wondered why he was not studying philosophy, thinking about the great ideas of science, pondering its methods and its creativity—which, he said, was as deep as any art—rather than actually doing experiments in the silence of some room, with some inscrutable instrument (the nuclear magnetic resonance spectrometer) clicking or whirring nearby. It was too lonely, perhaps, for someone who loved company, who shunned solitude, whose pencil-thinness bespoke a continuous motion, and whom I never once saw alone. Throughout his later years teaching college, it was the love of entropy—which he frequently equated with death—and the philosophy of science (what was the real meaning of entropy and its relationship to the one-way passage of time?) that lasted. The animation that in his phone calls today still exists reminds me of the younger man I knew in West Lafayette.

The town, west and north of the Wabash, seemed smaller than the university that sprawled in its midst. On the long walks to freshman lab, which I taught to support myself, I froze on the windswept plain that stretched to the west; but once there I enjoyed every moment in the warm sulfidic air, crowds of sleepy students opening their lockers, noisily fumbling with clamps and stoppers and notebooks as the dark morning turned light beyond the dusty windows. I remember in particular a student from that time, Rita, whom I dated later and whose delicate beauty became apparent when the lab goggles and apron were gone. She told me what it was like to live on the big farm her family owned not far away. It was another world beyond the blue-collar factory environment I had grown up in, a few short miles from Pittsburgh. Rita wanted to study medicine, become a family physician in a small town in Indiana, and make calls to remote farmhouses on winter nights. I have often wondered whether she lived out her dreams.

Curley's place was down the hill from campus, a greasy spoon unlikely to outlive old Curley himself, where I went to be alone, drink coffee, eat apple pie, and read Camus or someone Buckley or Harv had recommended, like Rainer Maria Rilke or the positivist philosopher Ernst Mach. I could not get enough of literature and philosophy in those days and maybe loved them as much as chemistry itself. The building I lived in had just opened and it was the highest building in town; from my window I could see glorious sunsets over the endless Midwest. In early evening there would be knocks on the door—a few friends would be off to dinner at the Memorial Union, with its colorful mural of steel factories and harvested fields that celebrated the land- grant virtues of hard practical labor that the university seemed to revere. Relatives and friends came to West Lafayette, usually by train in those days of smooth track and big stations and hot meals with heavy polished silverware that you held proudly in your hands—things that I thought would remain in this country forever. My brother, Eugene, would come up from the old Benedictine seminary in Saint Meinrad; my sister, Liz, from Pittsburgh; and Mary, my friend from high school, with whom I had gone to proms and dances and wide-screen showings of *Ben-Hur* and *North by Northwest*, came from Wichita, where she was studying, for a short stay.

The following year John F. Kennedy was shot, and it was in Edgell's classroom that I heard the news. I think Edgell was near tears when he told us, and I would see many tears in the days that followed. On the train ride home to Pittsburgh in the evening there were lights in farmhouse windows and in the small towns we passed, and you could feel the whole nation grieving, as though we were all one people. I had never known this kind of connection or felt its extent from one sea to the next like a trembling web that held us all in its embrace.

These are a few random memories of the place where I began to think of science in a new way. It was there that I began to realize there were human lives behind the equations I had been learning and the names I had been hearing all my years as a student—Kepler and Newton and Galileo and Lise Meitner and Lavoisier and Rutherford—who, until then, had been nothing to me but physical laws and maybe photos in some textbook, as lifeless as the paper on which they were printed.

❧

The undergraduate textbook I had carried with me to the university offered a glimpse into Boltzmann's work. You could hardly miss the beauty of it. At the age of twenty-four, scarcely older than myself, he had been thinking about the nature of gases, a subject that probably ranks for most of us somewhere close to dry toast on the scale of earthly excitement. But for the young Boltzmann, always given to looking behind the veil of phenomena ("strike behind the mask," as old Ahab had said), nothing could have been more worthy of his contemplation. What must a gas—nitrogen or oxygen or for that matter that improbable, extraordinary mixture, air—be like at the level of atoms and molecules to exhibit the properties that it does? Answering this abstruse question would be the great quest of his life, and the implications of that quest for us, for all of human history, have been immense.

But before I turned to the science, which I would only gradually come to understand, I wanted to understand the man. In West Lafayette, I gave this far more of my time than I should have. One of my tasks, beyond taking courses and presiding over the drowsy morning freshman laboratories and problem-solving seminars, wherein students not much younger than myself, newly arrived from the provinces of Indiana, came for a soupçon of human contact that was impossible to find in the impersonal lecture halls of the chemistry building, was to formulate a challenging research problem that would carry me swiftly through the years to my final reward, the PhD. On the subject of research I really had few original thoughts; every idea that came to mind had apparently occurred to someone else thirty years before and was now in moldering print in the esteemed *Journal of Something or Other* on our library shelves. Not a propitious beginning to a scientific career.

Instead of reading about the latest findings on europium Mössbauer or on the structure of the boron hydrides or on the thermodynamics of the hydrophobic interaction, I was reading about Boltzmann, picking up little anecdotes about his life from people in the department or from obituaries that had been written at the time of his death. (It was not until much later that fuller accounts of his life would be available.) I was beginning to think of him almost as a friend, someone I could enjoy a glass of wine or spend a cold morning sipping coffee down at Curley's place with. Meet my dear friend, the great classical physicist, Ludwig Boltzmann!

Since I was thinking of one day becoming a professor myself—indeed this was the only thing I could imagine now that I had forsaken the dream of professional baseball—I was especially interested in knowing how he had been regarded by his students. From the first, it seemed clear that he was anything but the model Austrian or German professor —austere, remote, obsessed with his own importance, disdainful of the lesser minds seated before him. In fact, he seemed more accessible than my own instructors, who were more concerned with reigning over their small research fiefdoms than in bringing knowledge to those who sat cowed and unquestioning in the cold amphitheaters they commanded.

As a young academic in the town of Graz, which then as now was the second largest city in Austria, Boltzmann was beginning to acquire a reputation for his passion and honesty, to say nothing of his utter brilliance. He was among the first to argue that women be allowed to attend lectures in physics and mathematics, and he frequently invited his research students to his home, where he told them about his own life and played them piano concertos of Bach and Mozart.

When he married Henriette von Aigentler in 1876, at the age of thirty-two, the couple built a house in the foothills above Graz, with a sweeping view of the village and the medieval university that lay in its midst. They had three daughters and, much later, a son. Boltzmann often held dances in the drawing room, where he played piano, to the great delight of his fortunate guests. Before he had even thought of physics, he had studied music as a child with the Austro-German romantic composer Anton Bruckner, later famous for his symphonies and choral music. Boltzmann would call upon his training and musical gifts all his life, including during his travels through America, when he once performed a Schubert sonata for Mrs. Randolph Hearst at her estate near Livermore, California.

Later, reading sketches of his life, I came to see that he was capable of a great range of emotions, which can be both blessing and curse. Once, crossing the Atlantic from Bremen to New York, he wrote of "the wonderfully rolling seas, each day different and each day more amazing," and at one point he wept at the color of the ocean. And standing on the deck at night he marveled at the phosphorescence and moonlight that played on the black surface of the sea. On the train east from California, he exclaimed at the beauty of snowcapped Mount Shasta, and on the great, clear lakes of the Sierra, reflecting pine and cloudless sky. Yet this deep appreciation for nature seemed almost eclipsed by the awe in which he held human courage and human achievement. The early explorers who mapped the far reaches of the earth; Pythagoras with his immortal equation; the visionary George Washington, "whose struggles had not only local but world historical significance"—all are acknowledged with gratitude in Boltzmann's marvelous and largely unknown travelogue, *A German Professor's Journey into El Dorado*, published a year before his death.

And he was no stranger to humor. In *El Dorado*, he wrote of a dinner at the Hearst estate in which he turned down several courses but reluctantly ate the oatmeal served as an appetizer. It was an unfortunate choice, and he concluded that even a flock of Viennese geese would have found this American delicacy more than distasteful. Passing through North Dakota, a dry state, on the train, he wrote with seeming delight about bribing the porters to get him good bottles of red wine.

But it was not his humor that most impressed his students, though they did write of how witty he was and how filled his talks were with amusing and unexpected asides. Mostly they were taken by the utter clarity of his lectures, their erudition, and by the staggering range of topics, from math to physics to philosophy, that he had mastered. One of his most famous students, Lise Meitner, who would be the first person to understand nuclear fission, said that Boltzmann's lectures were the most "beautiful and stimulating" that she had ever heard. She spoke of his enthusiasm and of how, in his presence, she felt "new worlds opening." Others were taken by the way each of his classes filled the largest lecture halls of the university, and by his humility, even at the height of his international fame. He could interrupt a minor mistake, a derivation, at the board by saying "that was stupid of me." In all of my education, until then and long afterwards, I could not remember one of my professors making such an admission.

Boltzmann's peripatetic academic career had begun in 1863 as an undergraduate at the University of Vienna, and had ended—after many detours—with his return from Leipzig to that same university in 1902. His inaugural lecture that year filled the largest hall, was attended by the Viennese press, and caused more than a little excitement. His introduction to that lecture summarized for me what he expected of his students and himself, what he expected of education and all that passed between professor and student. The man who believed in atoms, who believed in a fundamental world of particles underlying the beauty of mountain and sea, the plunging and rising ships of discovery, the sparkling lakes of the Sierra, the blueness of the sky itself, began that evening in Vienna on a much more personal note:

> *Today I only wanted to offer you something quite modest, admittedly for me all that I have: myself, my entire way of thinking and feeling. Likewise, I shall have to ask a number of things of you during the course of the lectures: strict attention, iron diligence, untiring will. But forgive me, if before going on, I ask you for something that is most important to me: your confidence, your sympathy, your love—in a word, the greatest thing you are able to give: yourself.*

Yet this man who was so unpretentious, humorous, and patient, whose personality projected a "radiant power" from behind the podium, was said by the Viennese physicist Franz Exner to be fundamentally unhappy. Even Lise Meitner spoke of his "mental instability," and Boltzmann himself, in a letter to one of his former students, Svante Arrhenius, said that he suffered from "neurasthenia," a poorly defined condition characterized by headaches, lassitude, muscle pain, and a host of other symptoms which perhaps resemble those associated with the more familiar fibromyalgia. Boltzmann sometimes jokingly ascribed his moodiness to the fact that he was born on the evening of February 20, between the dying dances and celebrations of Carnival Tuesday and the austerities of Ash Wednesday.

✺

The town of Duino lies near Trieste on the northeast coast of the Adriatic. In the early part of the last century it was a favorite summer destination for landlocked Austrians who needed a touch of the sea. In late August of 1906, Boltzmann and his family left Vienna to spend a little time there. The decision to go seems to have been Henriette's, since Ludwig had agreed to teach at the university that fall and felt the need to prepare lectures. Always there was that need, that nervousness, the deepening neurasthenia, perhaps. Even there amid the high cliff, with its castle, where once Rilke had gone to write the *Duino Elegies* in the damp stone rooms of the Princess Maria von Thurn und Taxis, there seemed to be no peace. And with him came that constant companion, that sense of failure, the recognition that maybe, as his great adversary Ernst Mach had said toward the end, he was the last who believed that behind the cliff, castle, and air were the unseen molecules and atoms he had argued for all his life

His fifteen-year-old daughter, Eva, discovered him. Disturbed by his long absence, she had gone back to the hotel to find him. While she and her mother were out swimming in the bay, he had hanged himself with a short cord from a window frame in their hotel. The papers in Vienna and the memorials at the university spoke of a man of brilliance "who had bestrode his time and his nation," but whom, as Franz Exner said, "envious fortune had denied inner peace." Meitner professed never to have understood, though she thought it might have been the depression, the "black dog" that had come to visit.

✺

Near the end of my first year, a lovely spring morning in Indiana, Rita and I were having breakfast at Curley's place and we were expecting Harv and Buckley to join us later for coffee. I had not yet found a project for my dissertation and was not even close. Rita told me she had enjoyed all of her science courses and was beginning to think she might even want to do research someday in physics or chemistry. Looking at her that morning, she reminded me a little of the young Lise Meitner.

I told her what I had been learning about Boltzmann's life and how I had come to see the creativity that lay behind his works, but also the struggles and doubts that had tortured him and, ultimately, the darkness that had forced his hand in Duino. He seemed to me now almost a martyr to truth, a secular saint.

Rita remarked how odd it sounded to her, especially after a year of chemistry, where atoms and molecules were treated as the furniture of the world, the tables and chairs of reality, that people in 1906—a mere fifty-eight years before our conversation—doubted the existence of atoms, mocked them, in fact, in prestigious gatherings, and considered Boltzmann a throwback to an earlier age. I could not have agreed with her more.

When Harv and Buckley arrived, they brought some positive news. Harv would be doing something in the laboratory of Norbert Muller, something with surfactants, and would be using the concept of entropy in his work. Boltzmann's concept, the very *S* that appears on the tombstone in Vienna. And with that, Buckley recalled some lines from the California poet Robinson Jeffers, who had written about the stonecutters and how, with marble, they challenged oblivion. I knew the lines, had encountered them as an undergraduate, and could even recite them. Jeffers knew that the stonework was "fore-defeated."

"Of course the stone will never last," Buckley laughed, "and Jeffers knew it. And Boltzmann, too. But that equation, well, that's another matter."

Buckley had made some progress in the search for a project, but he seemed more interested in what had happened at Duino.

"So," he asked, speaking in his resonant voice from behind a black beard that might have been down to his knees, "do you know the *Elegies*?"

I told him not really.

"Well," he said, "maybe these lines fit the story you've been telling Rita."

Since Buckley had a poem for all occasions, I was sure that in some way, possibly obscure to all but Buckley, they did.

"Please, Professor Buckley, tell us," Rita laughed, glancing up at Curley, who was smiling and pouring her a third cup of coffee.

And Buckley recited some lines from the first of Rilke's *Duino Elegies*, begun, also, in the midst of Rilke's terrible depression:

"Lovers, however, exhausted nature takes back into herself, as if there would not be the strength twice to achieve this."

We all looked at each other, a little perplexed. But the lines sounded right to me. Though I told Buckley I would have to give them much thought.

So the morning ended, the cash register rang the way they did back then, and from behind the counter Curley gave us his usual wave and his bright smile and thanked us for coming.

❧

I would be in West Lafayette only a few more months before I dropped out. I headed for California in winter over the dark soils of Kansas, down through west Texas, where the fibrous tumbleweeds hopped in storm across the hood of my Dodge Polaris, then up into the mountains near Santa Fe, when it was a town of seven thousand and the shawled Indians sat in the plaza near the Palace of the Governors with their bright beading and weaving and basketry, looking off into the deep valley of light. I had no idea where I was going, but I crossed the San Bernardino Mountains down into Riverside and Ontario, east of Los Angeles, into the country of orange groves where the air smelled of them, sweet and perfect, and got a job working the trees, with a canvas bag slung over my shoulder. And when that was done I worked on a ranch owned by a Hollywood studio musician, who said he didn't understand why I was there doing this and wondered if I had had a great disappointment back east.

It was like something my roommate, Durkin, had said when I shook hands with him for the last time: "I hope you find what you're looking for, Greenie." And I smiled at him and said thanks and told him how much I had enjoyed our time together.

Harv said, "I could never do what you're doing. Just give up on a PhD. Just walk away like that." Then he said, "I always thought you were looking for something. Some white whale, maybe. Moby Dick."

And I said, "I might have a better chance of finding him near the coast than here in Indiana." He smiled at that and I told him I'd write as soon as I got to wherever I was going.

It was a beautiful night for driving, and beyond the city lights, along the flat prairie, the sky was filled with stars.

Blacksburg
and Dallas

When I returned to West Lafayette from my travels in California, I stayed there for less than a year. While I was gone, my research advisor had decided to take a new position as chair of a growing chemistry department in Virginia. I decided to go with him. In early January, I drove the narrow fog-filled highways through the mountains and valleys from Pittsburgh down to Blacksburg. There had been a big storm and there was still snow on the ground but it had melted in the streets. The morning sky was blue, the air felt soft and warm, and I was happy that I had decided to leave Indiana and come here. There was again that feeling of newness and promise, and everything felt right.

Blacksburg was small and remote, lost in the southwestern corner of Virginia in the midst of the Jefferson National Forest. The houses where I looked for rooms had pictures of Robert E. Lee on the walls, and my landlady made rhubarb pie and said things like "Slow down, Yankee" when I talked. The few bookstores in town held large collections of Civil War literature, but not much else. Only a tiny place called Books, Strings, and Things, which was owned by a young couple who had come down from the East Village, had copies of Pynchon, Hesse, and Robinson Jeffers, and in the back of the place you could have bagels and chai tea and read *The Egyptian Book of the Dead*.

On my arrival, the plan was that I would work as a research assistant with a young professor by the name of Paul Field. He had just gotten a grant from the National Science Foundation to study a class of liquids called molten salts, which people at Oak Ridge Laboratories and elsewhere around the world were considering as possible heat transfer agents in a new generation of nuclear reactors. I had done no reading on molten salts before arriving in Blacksburg, and I thought of myself merely as a useful pair of hands.

It was not long, however, before this changed. I began to find the subject intriguing and to imagine what a molten salt must be. When you heated a white, seemingly featureless salt, say sodium chloride, sodium nitrate, or potassium nitrate, to a high enough temperature, it would begin to melt, to pool into a clear liquid, like water, only more syrupy, more viscous. You could imagine how the infinite array of ions, laid out in a perfect three-dimensional grid like some Bauhaus dream of order, would begin to tremble and shake as the temperature rose, and how, finally, at just the point of transition, the whole structure would fail. The charged ions would be cut from the bonded grid and sent to wander in the sweltering precincts of the liquid, freed, but still sensing the invisible chains of force that were all around. With the rising heat, the positive sodium, the negative nitrate would jostle and turn in an ever more frenzied randomness. It was as though you now had, in this ionic molten state, a pointillist drawing, a colorless Seurat, but one in which the dots that filled the paper actually moved. Not much was known about these strange liquids. And even less was known about how they would interact with gases, with nitrogen and argon and helium and neon and maybe even carbon dioxide.

I liked Field from the moment I met him. He had a no-nonsense manner and a wry sense of humor. I suspected he would be tolerant of my many quirks. He was a builder and a tinker and loved the laboratory and the challenges of making glass and steel and copper fittings into something that got you hard numbers. When I arrived, there wasn't much in the lab but a tall blue furnace, a vacuum pump, a small gas chromatograph, and sheaves of glass tubes, which would eventually become a transfer line for moving gas molecules around. With NFS money, Field purchased a great set of tools that reminded me of all our

*Overview of McMurdo Station. Ice-covered McMurdo Sound
is near the top of the photo.*

failed childhood rocket experiments and doomed astronomy projects, and I could only hope that a better fate with the practical arts awaited me this time. I decided to work with him and to make the chemistry of molten salts the subject of my PhD.

My three years in Blacksburg were not uneventful and, in a way, through happenstance and accident, they led me to what I had been looking for. In Bern, Einstein had shown that a dust particle in water is ruled by randomness. And the path of a nitrate ion in a molten salt is a thousand times broken, governed as it is by the unexpected collision, the improbable glance, the fits and starts of molecular motion. Looking back, I can see how these vagaries applied in a similar way to my own life, and maybe, to a greater or lesser degree, to every life, even though I was reluctant to admit it. In retrospect, the element of chance was nowhere clearer to me than it was in Blacksburg, Virginia.

❀

It took me six months or so to learn glassblowing. In my hands, it was by no means a fine art, but I became proficient at it and could fashion right-angle bends and T-joints and weld in stopcocks and mercury manometers to what was becoming a functioning twelve-foot sculpture along the gray walls of the laboratory. In the early morning the sun would rise over the peaks of the Jefferson National Forest and turn the glass to silver.

Within a year the apparatus was completed. The furnace was attached, a stainless-steel pot that contained the salts was enclosed in the furnace chamber, the circulating pumps were set going, and when the crystalline salt had turned to liquid I began my experiments. From my desk, I could hear the gas gently bubbling through the melt. To me, it was a tiny symphony, and I knew I would soon have data, and hopefully a dissertation.

The fall of my second year, I met a girl at a nearby college. We spent whatever days we could together. There were picnics at Claytor Lake and walks in the newly fallen leaves. As the days became shorter and the air crisper, we became more serious. I visited her family in McLean, near Washington, and she came to Pittsburgh for long weekends. We decided to marry and to plan a formal wedding for June. In the end it was all a mistake, and by the following autumn our short marriage was over.

When I moved out of our apartment, I had not arranged for a place to live. For weeks I slept in the backseat of my old Polaris. I began to look like the men I had seen under bridges in Long Beach and Los Angeles. I worked in the lab at night and tried to sleep during the day. One morning, one of Field's students asked me straight out where I was living. I told him I was sleeping in my car. He said, "You know, Roger Hatcher and John Hall have a trailer outside of town. They're looking for a third person to share the place. Are you interested?"

I had no idea who they were, but he told me they were both biologists and were friends of his. So I said, "Sure, why not." We made arrangements and that week I moved in. I had a tiny room with no window, but the place had a shower and a well-stocked kitchen and a television where we watched the evening news together and talked about the war. Hall had a large tape deck the size of a suitcase and a complete recording of *Sgt. Pepper*, which in short order I committed to memory.

It wasn't long before I decided this was a good living arrangement. Hatcher was industrious, witty, an outgoing entrepreneur. The trailer he rented supplemented his income from the biology department, and he was determined to finish his PhD in another two years. Hall had been to Vietnam and was just a few months away for completing his master's degree in microbiology. I think Hall believed I was a communist or at least a member of SDS, and he would often joke about our trailer being under FBI surveillance. He had worked with Robert Benoit and had done his research in a place I had never heard of—the McMurdo Dry Valleys of Antarctica.

One evening in the spring of that year, we were having dinner, the aluminum door of the trailer wide open to let in a breeze that came down from the meadows and the distant hills. We had just finished listening to Eric Sevareid's wise commentary on the war when Hatcher turned to me and said that Benoit had been asking about me. "He wants to know if you'd be interested in doing some chemistry down on the ice," Hatcher said. The question came as a shock. The spring day, the CBS news, the meal I had prepared all faded.

I was in the midst of finishing the gas solubility work and was looking forward to writing up my results and maybe getting them published. But I had seen enough of Hall's slides and listened to enough stories of the exotic places he had been that I wanted to know more. "You should do it, Green," Hall said. "It's a once-in-a-lifetime chance. And I know you'll like the navy people at McMurdo Station." He gave me a big ironic smile and turned toward the TV.

That night I thought about it. I could use a break from the lab before I began to write. The next morning I told Hatcher I'd meet with Benoit and see what he had to say. The following afternoon I went to Benoit's office. Maps of Antarctica and reprints of importtant papers were scattered around. Benoit was tall and rugged-looking, and he spoke with enthusiasm about his project, about the odd lakes he was studying, about what he hoped to accomplish. He was a pioneer, the first scientist to study bacterial colonies in these distant waters. Benoit and Hatcher were leaving for Antarctica in August, on the first flight south. He himself would have to return home in October to teach, but he wanted his students to stay until Christmas. He needed a chemist, he said, and wondered if I would be interested.

I spoke with Field as soon as I could to get his advice. To my great surprise, he was all in favor of my taking Benoit's offer. "After all," he said, "you have no other responsibilities now, except to this." He swept his hand through the air to frame the solubility equipment, our lovely "glass sculpture." In the momentary silence, I could hear the argon gas bubbling through the lithium nitrate melt. I decided to go.

Physical chemists use a concept that comes from the molecular vision of Boltzmann. It is called the *mean free path*, and it refers to the distance that a molecule travels freely in space before it collides with another. These distances are usually small. The collisions themselves can be of various kinds. Sometimes they are just glancing, with one molecule barely grazing another, so that the trajectories of both are hardly changed. On the other hand, some collisions are direct, head-on, and violent, and in these cases, depending on the molecules involved, a chemical reaction might take place. You can imagine two molecules— hydrogen, say, and iodine—coming toward one another at some outrageous speed and, when they meet, sundering, so that one of the hydrogen atoms and one of the iodine atoms go off together as a new molecule, HI, or hydrogen iodide. This happens so suddenly, this exchange of partners, and so perfectly, with each electron seemingly knowing where it is to go, that you just have to shake your head in amazement. The violence of these collisions, of course, depends on the temperature, as does the rate at which molecules react to give new molecules. So frequent are molecular collisions and so short are their mean free paths that a single molecule might travel the distance from Boston to San Francisco just to move three feet.

I like to think that this concept, in some rough and imperfect way, is a metaphor for what happened in Blacksburg. Actually, for what happens anywhere, to any of us. I had never thought of Antarctica. I was on my way to doing a PhD in high-temperature physical chemistry. I was thinking I might find a position at Oak Ridge or in some university chemistry department where I could continue the studies I had begun in Blacksburg. But then an improbable cascade of events occurred: marriage, divorce, a chance meeting in a laboratory, a vacancy in a trailer court, the happenstance of Hall, Hatcher, and Benoit, set in place like the fielders in a triple play, and, finally, the lonely austere valleys by the Ross Sea, with their lakes and glaciers and streams, something out of legend and myth, the curtain of my life slowly rising.

What Blacksburg revealed was the randomness at the heart of it all. Somehow that had to be taken into account. Boltzmann had seen that randomness all along, had seen chance and probability at work even in the second law. And, looking back, we can see it, tragically, in his own life. In the vacation at Duino. In the silence in the non-existent connection, the cut wire between himself and Einstein.

⚛

It was August 1968. We drove to Washington and left from Andrews Air Force Base on August 27. The plane was the C-130 Hercules. The flight took us first to Travis, near San Francisco, then on to Hickam Field in Honolulu. We stayed for two days in the city, when it was small, an island village, really, before the high-rises went up, and the "pink hotel" was the largest place in Waikiki. We stopped in Samoa and then flew to Auckland and, finally, to Christchurch, farther south.

Days later, when we got to Antarctica, the continent was still locked in darkness. Six months of night was beginning its slow lift toward daylight. It was bitter cold when we landed. The strong winds tore at the Herc. I staggered down the high stairs from the plane under the weight of my parka and the orange canvas bags we all carried. The snow was dry and it crunched under my boots.

The cindered roads smelled of diesel. The buildings were warm and smelled of coffee and chocolate. The white volcano, when it hove into view, rose thirteen thousand feet, and the tiny village of McMurdo —like something under Popocatapetl—lay at its base. The sea ice, where the Scotts' ships had once anchored, was solid far off into the north, and the white peaks of the Society Range lay in near-darkness to the west. You could still see the moon and the southern stars. Antarctica looked as big and dead as a Jovian satellite when you stood on a prominence in the silence. You thought you were the last person. But then in the officers club, you could hear the sound of popcorn being made, where they served up "black Russians" and "lion browns" for a quarter, and the small seats of the movie theater were torn from use like they were when I was a kid. It felt like it was home even when the building shook from the wind and the projector, with its loops of tape, tangled and died. I had not been this happy for a long time.

We did the limnology we had come to do. But it was odd being a chemist among the people who studied the big fish and their survival in water that should have frozen their blood solid, or the seal people from Scripps, with their wooden houses out on the sea ice. I used to go there at night by the deep hole where seals appeared, their eerie breathing from the big head, alive with me, their brown eyes searching my face and then submerging and returning to look again. The guys who dove for starfish in the Sound were from Scripps too. They lived with the bright starfish in the dark muck of the sediment. When they got their photos they showed them to everyone in the biolab on Sunday nights.

Science in the field, I soon learned, was different. It was harder in a purely physical way. I remember the cutting wind, the great mountains all around, the cold stream when I dipped in it in the morning for water. Every measurement held a memory of its taking, the labor of its taking. Some of the lakes were smooth and the sleds were easy to pull, but others were rough. We had to unload the sleds and lift them and load them again. The days on the lakes were eight to twelve hours long. Sometimes I didn't want to stop and I came back and collapsed on the shore and fell asleep right there with the hood of my parka up. When the choppers came and moved us to another lake, there were five hundred pounds of gear to load, maybe more. Tents, generators, drills, and water samplers all had to be wedged into the Huey and then unloaded and set up again. For energy, I lived on Cadbury bars and peanut butter and drank hot chocolate near the Coleman stove. In the field, I felt the glorious exertions of my youth. Sometimes it felt like pure play, even though I was doing science. And I just had to look up to see the beauty all around me. The mountains disappeared in snow and reappeared, transformed in whiteness and light, the way the hills of Pittsburgh once did. Back at McMurdo, they had a soft-serve ice cream machine that anyone could use. I filled cups and soup bowls with it after every meal.

In the Dry Valleys, I had so many questions. Why were we finding so much dissolved oxygen in the water? I had expected there would be none. Maybe our methods were wrong; maybe one of the reagents had gone bad. After all, the water was under twelve feet of ice. It was no longer in contact with the atmosphere. And why did the ice, which was so perfectly blue, have these long columns of bubbles embedded in it? And why did the deep waters in some lakes smell of sulfide, while others did not? I never answered these questions, not that year. Nor did I know the chemistry of the rivers that flowed in the austral spring when, for a few weeks, it was warm enough.

※

It surprises me now. But in time, I rarely thought of Antarctica. Maybe it was just a once-in-a-lifetime event, as Hall had said. I came to believe that. I finished my PhD and thanked Field for all he had done. He had been a great adviser and I had learned my way around the laboratory, and our work would soon be in print. When I left, I wanted to try something new, so in 1969–1970 I accepted an appointment in Belfast, Northern Ireland, and began a study of polymers. But in the spring of that year, my father died suddenly and I returned home to be with my mother. In Pittsburgh, I took a job studying air pollutants from printing presses. We traveled everywhere in the East and Midwest, and I sampled smokestacks in Chicago and Detroit and Philadelphia and gave lectures on air quality. At night I studied philosophy.

Later, I took another postdoc. We worked on the nature of water, its structure and bonding and how it reacts when a molecule or an ion is dissolved in it. I continued to study philosophy and thought I might do another PhD.

It was not long before Wanda and I married. She had known my sister for years and had lived directly across the street from us. We could talk and laugh forever, it seemed, and she told me this was just the beginning of a conversation that would last a lifetime. I was skeptical, but in the end she was right.

Two years after we were married, I got an offer to teach in Ohio. In my first year at the college, our daughter was born and I was, improbably, a father. I was learning how to teach and how physics, chemistry, and philosophy fit together. I taught chemistry courses, too, on the oceans and air quality and Antarctic lakes—things that I had come to know firsthand over the years. To my surprise, there was an opportunity to do research, but I had no idea what I wanted to do. I looked for a mentor and found one in Texas.

In the summer of 1976, Wanda and Dana and I went to Dallas. We lived in a small apartment outside the city, in the once-empty fields near Plano. I worked for two months with G. Fred Lee, a renowned expert in limnology and water chemistry and a professor of environmental science. G. Fred was a large man and I thought of him as Wallace Stevens's "Emperor of Ice Cream": "the roller of big cigars, the muscular one," the one who "whipped in kitchen cups concupiscent curds." We had corresponded for two months, but I met him for the first time at his ranch on the prairie outside Dallas. On a hundred-degree afternoon, he showed me the horses, the tennis court, the swimming pool, and the large office and laboratory complex where his secretaries and graduate students and postdocs all worked. "I could never afford all this if it weren't for the grants and consulting," he said, with obvious pride. Then he asked me to join him and his wife, May Beth, for a drink. He told me he had grown up in the orange-grove country of California and that he once played guitar with a country-and-western group. He had been poor, but he was good at science and knew he could make a living at it. In the late fifties, he got an appointment at Harvard to do his PhD with the Swiss aquatic chemist Werner Stumm, whose practical and theoretical work on Lake Geneva had come to define a whole new field of knowledge. When the environmental movement came along, G. Fred Lee was ready.

Lee was in his mid-forties, a man of big projects and big plans. He was studying Lake Ray Hubbard, the huge new impoundment of the Trinity River in Dallas. He was consulting with the governments of Spain and Italy on eutrophication, working with the Army Corps of Engineers on Port Aransas, Galveston, and other dredging sites around the country. He was studying wells and hazardous waste disposal and looking at groundwater contamination. He had a grant with the Environmental Protection Agency, in which things were not going as well as he had hoped. He wanted me to take a look at the equipment to see if I could help.

The project involved the movement of organic solvents through clay liners in waste disposal pits. It was feared that the solvents could break through the protective clay and contaminate surrounding groundwater. I knew groundwater contamination, once it occurred, was more or less forever. So I got involved, even though I was reluctant to work with benzene and trichloroethylene—solvents people were just beginning to understand might cause cancer. But I was young and immortal and, besides, like every student of chemistry, I had been in contact with benzene a thousand times. I found the work interesting, and we published some well-received papers.

Toward the end of summer, Lee and May Beth and Wanda and I went to dinner at a fine restaurant in Dallas. Fred asked me if I would work with him full time. "I can pay you more than you're making in Ohio," he said, "and I know you enjoy this research as much as I do." I looked over at Wanda and saw a skeptical frown on her face flicker for just an instant. "I don't know," I said. "I'll need to think about it."

In the lounge, we talked about a lot of things, but Fred seemed fascinated by Antarctica. He said he had known some of the early researchers, like Gene Likens, who had worked on Lake Vanda in the early sixties. But he had not kept up with the research. I told him that not much had been done in recent years. Then he asked if I had ever thought of going back. "You could study trace metals," he said. "They're a big issue today. Or phosphorus inputs, eutrophication. I think the National Science Foundation might give you some funding."

The soil was fertile and Lee had dropped a seed. After we returned to the apartment and settled with the babysitter, I talked to Wanda about Fred's offer and his comments. She told me she had no interest in coming back to Dallas. Ever. But she thought the idea of a proposal to study trace metals was a good one. We had just bought a new house and she was feeling comfortable in Oxford and thought an NSF grant might ensure tenure. It had been eight years since I had last seen Antarctica, and I had thought of it only sporadically since that time. But something had happened there at the restaurant with Fred—all the old questions started to return, and with them came the images I had filed away for so long. In the gloom of our crowded little place, with our daughter peacefully asleep, I knew Dallas was over; but I felt that something new was just beginning.

✾

Images from childhood, so many of them turned outward toward the stars and planets and the deep mysteries beyond. They entered and unlocked whatever held me in the fastness of place and they gave me the long-evolving universe of brilliance and darkness that Hubble must have seen there on his cold mountain. Newton had said we are children at the edge of a sea we cannot fathom, and on clear winter nights, the Dipper rising over the rising hill where we played, I had no words for it. I wanted to touch what was out there, use the rocketry of the back porch, the propulsion of gunpowder, the rings of Saturn floating in Bobby Newport's lens, to find my way there.

And there were other images that turned inward, that took me into what was small and so distant in its smallness, that let me wander among interstices and spaces around the sudden curve of atoms, into the swarming electrons of Bohr and Schrödinger, the tiny swans at the heart of matter, who in their flight know no rest. And the dust mote, tugged and pulled, the battering tripartite molecules we cannot see, as if a school of minnows, diaphanes of jelly, could move a carrier side to side. The glowing ember in the furnace, a red sun dying before my eyes, turning fugacious, turning to gas, in my memory lives on, dies each night in its beauty. I held on to these as though in their presence long ago my heart really had opened. Were not the molten salts, the ions, positive and negative, jostling in the night-glow of the lab furnace, the same enactment of the trembling, fleeting miniature, the same liberating release into all that was strange and hidden?

Science was a world of dream, far from classroom and lecture hall, from the clay fields whose chalk lines ran forever, from the black-and-white of the living room TV and the warmth of evening meals. At night I felt a joining of myself to all that was—the star and the hand before my face, the fallen leaves and my own breath, the sea and what moved in it and my own coursing blood—and as I learned more the mystery only deepened, the connections became a skein of magic. Science was the mountain on which I could stand, the pure air of which Einstein had spoken. It was the geometry of Kepler, the spheres whispering far above the tumult, the peace of heavens. It was the spectrum of Bohr—red shading to violet, the four crisp lines—that spoke of logic and reason beneath the clear splendor. It was all of these things and wonderment that kept me knowing, that beneath our sun and beyond, it would always be new. Who could want anything but more of this?

After Dallas, it was another three years before I got the first grant. A thousand times over and more, I thought, Boltzmann's molecular randomness was surely at work. The probability of that phone call coming from NSF, when reckoned from the winter day I first arrived in Blacksburg, was so small as to be nearly zero. As much as I didn't want to believe it, maybe it was true that, as Mlodinow had said: "Randomness rules our lives."

Los Angeles
and Prague

In the summer of 2009, our older daughter, Dana, had major roles in two plays at the Old Globe Theatre in San Diego. We spent lovely evenings in Balboa Park beneath the stars, marveling at how accomplished she had become as an actress. I spent a morning at Scripps, looking for colleagues I had met at McMurdo in 1968. I listened to stories about Charles David Keeling, the pioneer who had first begun measurements of carbon dioxide in the earth's atmosphere.

Next we drove north to Santa Monica to spend a few days with Katy. One afternoon the four of us went to the Griffith Observatory, a place I had not expected to become as important to me as it did. Memory glides over time as easily as a stone over ice. It is free to move across the latitudes and longitudes of that dimension with the slightest of forces. At the Griffith that afternoon, my mind slipped back to 1995, a year that included not only Antarctica but also the beginnings of my historical explorations.

I had spent a year as a postdoc in Ireland, and on various occasions had visited the Continent and England, but in none of these travels did I think much of the history of science, not even of Boltzmann or of the Cambridge of Alan Clifford. But by 1995, perhaps as a result of having taught a course on paradigms and revolutions, with my friend and colleague Barbara, I had given much thought to the unfolding of five hundred years of scientific knowledge. In 1995, it was as though some part of me had awakened from a very deep sleep.

When I recall that year, I recall it not as it was actually lived by me —the trains (always the trains) from here to there, across this river and through that mountain pass, and into that city, with its little shops and pastries and motionless mimes—but as we had taught it, beginning with the great astronomers who had given us position in a universe of stars and had set just the right problems to open up the science of physics and then, beyond that, chemistry and geology. To the extent that I could, I was retracing history, retracing footsteps in the thought and work of Copernicus, Kepler, Galileo, Newton, and Einstein. My journey, of course, did include the tomb of Ludwig Boltzmann, at one time in my life the sole objective of any trip I might take to Europe. His gravesite, sad and forgotten as it was, offered more than I could ever have expected.

✺

The Griffith Observatory is in the Hollywood Hills high above the restaurants and shops of Los Feliz. Standing at the entrance to the park grounds, before the observatory domes, there is an alabaster sculpture completed in the 1930s in honor of the great astronomers. Among the thin monkish figures arranged in a circle around the central obelisk are the grand architects and authors of modern astronomy: Copernicus, Kepler, Galileo, and Newton.

In these names there was hardly the resonant beat of chemistry. None of these explorers of the heavens, of the planets and stars and the paths of comets, had ever heard of ions or molecules, the jostlings of sodium and nitrate in the midnight furnaces of Virginia. Nor had they ever imagined an atmosphere of nitrogen and oxygen admixed with a trace of carbon enough to warm their days. They had not known of a land far to the south held in darkness and bitter cold, where warmth and sunlight came slowly, and for a few precious weeks sent water in cascades and brightness along the thin ribbony channels into lakes that had been scoured with winter wind and sand. The lakes were jewels in valleys of stone, and the valleys were mere dots in a

*The Astronomers Monument at the Griffith Observatory
in Los Angeles.*

landscape of ice and more ice, an encirclement of ice bearing down.
Who in those centuries could even have imagined it—the vast ice fields
and glaciers, the threatened ozone that lay high above, the molecules
and ions that swam in the streams and in the lakes. And yet for me,
so much of wonder started with them, with those figures, with the
stars and planets that haunted their dreams. Without them, there
could have been no Boltzmann or Bohr, no Meitner exploring the
pierced shards of the nucleus; the modern world we call home would

have been some shrouded land beyond ice and mist like a continent impossible to descry.

On his deathbed in 1543, Nicolas Copernicus offered the world a new vision of the heavens. It was given in the form of a book, *The Revolution of the Heavenly Spheres.* In its pages, Copernicus proposed the idea that the earth moves about the sun in a period of 365 days and that each day, through darkness and light, it turns on its axis. This single eversion, placing the sun, instead of the earth, at the center of the planets, and giving the earth a spin about its own axis, can be seen from the vantage point of our own time as one of the great transformations in human history. It was the beginning of the Scientific Revolution and, in a sense, it was the genesis of the modern world.

In the rotunda of the Griffith, a Foucault pendulum swung from a bearing forty feet above in the ceiling. A crowd had assembled around the deep well—children, parents, grandparents—watching the to and fro of the massive bob as it sliced a path through the air. Every ten minutes it took down a small peg on the floor as the crowd peered over the railing in expectation. It was loud in the hall; people were waiting for the next peg to drop. The curator was saying, "This is the closest thing to perpetual motion that we have. At this latitude, it takes forty-two hours for all of the pegs to be knocked down, for the Earth to complete a single rotation." Someone stepped for the first time into the circle of onlookers and said, in surprise, "Why are all these people looking at this?"

Later, Wanda, Dana, Katy, and I went to the sky show. It was the usual thing: the total darkness, the disembodied voice, the sense that you were seated on a high hill in some distant windless place with the stars and the full dust of nebulae slowly rising. Aside from the godlike voice, for a short time I was the only one in this virtual world.

Long before I had the words, before I knew the names, the night sky had lifted me in some kind of gauzy trance. Ptolemy, the great

Egyptian astronomer, once wrote, "When I gaze upon the stars, I no longer walk on Earth," and Rilke said, when writing of the stars, "everything is its own sigh at being what it is." I had this sense, too, of joining and being joined, of being one small thing in the milk river that flowed infinite above me and around me and that sighed at being what it was, that marveled at its own being. In the heavens, at an early age, I first felt, intuited, what must be the beauty in science—not just this one science, not just astronomy, but any science that gave you entrance into the mystery and heart of the world.

Many years earlier, in Ohio, the four of us drove a short distance out of town and pulled to the side of the road. No one was around. The air was chilly and the corn was so high you could walk into it and get lost if you wanted. We all sat on the hood of the car and looked off to the sky in the west, off toward Indiana. Wanda and I knew that in our lifetimes we would never see Halley, but that night from the cornfields, just south of Acton Lake, comet Hale-Bopp was large, slow, and stately in its ice-shedding passage. You could see the thing moving. It was so bright there could be no doubt what it was. It was the first comet any of us had ever seen.

Outside the Griffith, we looked down through the sere folding of the Hollywood Hills. The famous sign was there, and off to the left was the city of Los Angeles. The paved roads, the cars, the tall buildings, the lights and studios—the whole modern panoply of the city—connected back to the figures on the pedestal out front—to Copernicus, Kepler, Galileo, and Newton, the four who drove the great revolution that led to all of this. You can draw a nearly straight line through history, from them to all the visionaries who followed, who powered our engines and brought us light. From them came the instruments and methods that extend our lives, that sire the rovers that inch across Mars in its russet dawn with signals of hope. They gave us Los Angeles, Singapore, and New York beaded in radiance across the midnight globe. And they gave us all of the smallest things that exist beneath our seeing: the jostling, wandering atoms that Boltzmann knew were there; the orbitals and the spun clouds of hydrogen and oxygen; the frames of carbon built on carbon that lay in pools in the darkness beneath Titusville, Pennsylvania, and beneath the sands of the Arabian Peninsula.

On the way out of the park, in front of the Astronomer's Monument, there was a kid gleefully hopping around the sun, hopping along the Copernican path of Saturn inscribed in the ground.

᠅

Fourteen years earlier I was in the city of Prague, walking in the old square, when the heavens opened. People scattered into every shop and doorway they could find. I ran into a church I had been admiring from the outside. It was the Cathedral of Our Lady before Tyn. Older by two centuries than Copernicus's great book, this late Gothic church, with its castellated towers, dominates the square of the old town. Inside there is the tomb of the astronomer Tycho Brahe, whose naked-eye astronomy would eventually help buttress the Copernican system. Brahe, with his gold-and-silver nose, is buried in the Cathedral.

It may seem to us, after years of seeing our spinning globe trace its path around the Sun—not literally, of course, but in countless classroom and planetarium demonstrations, from childhood on—that there could never have been a question about what Copernicus wrote in 1543. But as soon as the book was published and its tenets grasped, criticism came from many directions. The theological considerations alone were troubling. As Martin Luther said, "This fool wants to turn the whole art of astronomy upside down. But as the Holy Scripture testifies, Joshua ordered the sun to stand still, not the earth!" For many, it was disconcerting to think that God could have given the privileged center of the universe to anything but our own planet.

More seriously, perhaps, there was the empirical problem of parallax. If, in fact, the Earth moved around the Sun, it was expected that there would be some displacement in the position of the fixed stars as seen six months apart. Your finger appears to shift if you look at it first with one eye closed, then with the other. So, too, should the North Star change its apparent position in the sky, depending on the angle of your vision. But this was not the case. There was no shift, at least none that anyone at that time could measure. Finer instruments were needed.

There was also the everyday, commonsense experience of feeling the Earth to be completely and forever motionless rather than the spinning, onrushing, circling object represented in the Copernican system. How could we be moving when our senses tell us only of stability? As one sixteenth-century poet wrote with memorable brevity: "Never should an arrow shot upright / In the same place upon the shooter light." And we can think of many similar examples, drawn from our own daily observations, which argue in favor of the long-established Earth-centered world of Aristotle and Ptolemy. If Copernicus's theory was to gain traction, it would require genius to lend it credence. And this genius came in the persons represented at the Griffith.

When I teach this material, I say to my students: "Let's assume it's 1543. We have all read *De Revolutionibus*. We have had a chance to think about things for a few weeks. Then someone asks, 'Well, what are you, a Copernican or an Aristotelian?'" I ask them to cast a secret ballot, just in case they are a little nervous about the Church. After all, who wants to be tortured? The vote usually comes out about ten to one. They are Aristotelians.

※

I could still smell the wax from the candles as I walked from the church into the square. The rain had stopped and the morning had turned sunny. I decided to have coffee within view of the Astronomical Clock across the square that Johannes Kepler is said to have worked on.

Many writers and historians speak of Kepler with awe. He was a mystic, an astrologer, and one of the most profound thinkers in the history of science. I remember an article I read once in a seminar that Larry Laudan taught, entitled "The Copernican Disturbance and the Keplerian Revolution," written by one of my favorite philosophers, the subtle and inventive Norwood Russell Hanson. Hanson had had a life that seemed Keplerian in its scope and intensity. He had studied trumpet with William Vacchiano, whose students at Julliard included Miles Davis and Wynton Marsalis, and had played at Carnegie Hall. He had been a Marine Corps pilot in World War Two and was known for his stunt flying, most famously his loop of the Golden Gate Bridge. As a philosopher and historian, Hanson had taught at Cambridge, Indiana University, Princeton, and Yale, and had argued that the history and philosophy of science should be combined into a single discipline. His article on Kepler was published in 1961, six years before Hanson's plane went down as he was flying through dense fog en route to Ithaca, New York. He was forty-three.

We think of the Scientific Revolution as having begun with Copernicus. We even give it a precise date, 1543, and this date is largely accepted by historians. But there is something about Hanson's argument that I have always liked. He claimed that the real revolutionary was Johannes Kepler, for twelve years a resident of Prague. Copernicus had shifted the point for astronomical calculation from near the Earth's center to near the center of the Sun, something that might have been done far earlier, even in Ptolemy's time, with the data available. But Kepler's contribution was the radical shift away from the circle—once deemed the perfect and only acceptable path for a heavenly body to follow. It was a shift to another geometrical figure: the ellipse. With the ellipse, Kepler fashioned his three laws of planetary motion, laws that we use today in the placement of satellites. It is Kepler, not Copernicus, who forms the backdrop for the astronomical calculations in Newton's *Principia*.

Kepler came to Prague in 1600, fifty-seven years after the death of Copernicus. He came both by accident and design. Accident because near the end of the century, the Archduke Ferdinand, in the process of attempting to rid Catholic Austria of the Luthern Heresy, had closed Kepler's school and ordered all teachers to leave under threat of death. Expelled from Graz—where two centuries later Ludwig Boltzmann himself would spend the happiest years of his troubled life—Kepler sought out Tycho Brahe, the one man in Europe whose astronomical measurements could be counted on to help him develop his new theory of the heavens. In part, that theory had arisen out of a deep faith in numbers and geometry, and in the belief that the cosmos must be based on the regular solids of Plato.

His *Mysterium Cosmographicum,* published before the Prague years, was a curious construction of the six known planetary spheres—those of Mercury, Venus, Earth, Mars, Jupiter, and Saturn—inscribed about the five regular solids: octahedron (innermost), icosohedron, dodecahedron, tetrahedron, and cube. The sizes of the Copernican orbits were known, so Kepler could arrange the planets to proper scale. It was, if nothing else, an ingenious consilience between the ancient geometry of the Greeks and the new astronomy. "God is geometry," Kepler had said, and he believed in his bones he had unlocked the secret of the cosmos.

Kepler's collaborator, Brahe, the famed naked-eye observer of the heavens, had recently come from Denmark at the invitation of Emperor Rudolph II. Brahe was a man of wealth and position and of grand and legendary appetites, a libertine. He had lost his nose in a college duel—fought in the dark, no less—over whether he or his adversary was the better mathematician. His associates at his Danish castle included a clairvoyant jester, who frequently sat beneath his dinner table, and a pet moose, who unfortunately drank too much beer one festive evening, tumbled down the stairs, and died.

Uraniborg, Brahe's former castle and observatory off the coast of Denmark, was a vast baroque structure of domes, porticos, and decks suspended in air, from which he could look out in all directions to the planets and stars. It was furnished with quadrants of gilt brass and carefully crafted sextants. One of his mural quadrants was nearly twelve feet in diameter. Kepler once remarked that any one of Brahe's instruments would be worth his (Kepler's) entire fortune and that of his extended family. These were the finest and most precise instruments ever made for pretelescopic astronomy, and Brahe's unrivalled data sets would prove to be a massive treasure for Kepler.

Brahe was convinced that the future of astronomy lay in experimentation and in accurate and precise measurement. He narrowed the traditional precision of observation, which had been about ten arc minutes, down to two. He was the first to observe the supernova in the constellation of Cassiopeia (1572); and from the parapets of his castle above the sea he tracked the course of the comet of 1577. He wrote of all of these in manuscripts, printed on his own printing press, and through these articles challenged the Aristotelian doctrine of the constancy of the heavens. The star referred to in *Hamlet* is almost certainly Brahe's supernova of 1572. ("Last night of all, / When yon same star that's westward from the pole / Had made his course to illume that part of heaven. . . .")

Brahe is remembered today as the best of Europe's observers, but not as a theoretician. The world that he fashioned from his readings of the planets and stars still had the Earth as its center. About this focal point, the Sun, the Moon, and fixed stars moved, and about the Sun, the planets Mercury, Venus, Mars, Jupiter, and Saturn revolved. In a sense this construction was a compromise between Aristotle and Copernicus, and its lifetime would be short.

Kepler recognized Brahe's work to be the kind of precise astronomy that he could only have imagined in Graz, in his own observatory of sticks and strings hung from ceilings. He immersed himself in the data, let it serve as a guide and constraint upon his own imagination. The problem of Mars, which he had been assigned by Brahe, became an obsession. And Brahe's data on Mars, which limned its path through the heavens, including its bizarre retrogrades, could simply not be explained using the inviolable circle that had been a fundament of Aristotle's, Ptolemy's, and Copernicus's systems.

It was Kepler's struggle with Mars that gave birth to his new astronomy. Mars, as he tells his readers in the second of his major works, *Astronomia Nova*, "is the mighty victor over human inquisitiveness, who made a mockery of all the stratagems of astronomers, wrecked their tools, defeated their hosts; thus, did he keep the secret of his rule safe throughout all past centuries and pursued his course in unrestrained freedom." Even after four years of calculation, the disagreement with a circular Copernican orbit involved a tiny eight minutes of arc, but these eight minutes were far greater than the uncertainty associated with Tycho's measurements.

※

Late morning I was restless. My mind was wandering with associations, as it often does, like some Billy Pilgrim unstuck in time. I walked over the Charles Bridge, which was ancient, flat, and low across the river. I bought a watercolor from a street-corner artist that showed the scene I was looking at, the turreted castle up on the hill, the gray, striated sky behind. I went to the Franz Kafka Museum thinking that this very scene must have played out in Kafka's mind as he wrote *The Castle*. (There is a strangeness to this city that suggests Kepler's age, evoking thoughts of witchcraft and burnings, divination and horoscopes, cast even by Kepler himself.) I went to the square again for coffee, past the astronomical clock, and took particular note of its skeletons and its dark reminders of time and death.

I was first told at the Buhl what had happened here in Prague in those extraordinary twelve years. The summer before my junior year in high school, a friend I had known from my days delivering papers told me about a course in astronomy that was being given at the Buhl Planetarium, on the north side of Pittsburgh. The course would last only a few weeks, so I decided I would go there with Charlie to see what I could learn. The Buhl had long been a favorite of mine, with its Zeiss projector that rose from beneath the floor like an ancient sea monster to cast its stars on the theater dome. At Christmas, there was a model train layout with mountains and lighted towns that were threaded through by a thin line of gondolas and boxcars and silvery trains whose passengers were silohouetted in tiny windows. In the foyer, the Buhl had its own Foucault pendulum. In a special room, you could see a full-size replica of Brahe's mural sextant, which appeared to me, even then, to be a work of art.

The first law of Kepler became the familiar statement of the Buhl's lecturer and of textbooks everywhere: *Planets move on elliptical paths, with the Sun at one focus of the ellipse*. It first appeared in *Astronomia Nova* in 1609, and concerned the path of the planet Mars. It is written almost as an admission of failure. Kepler says, "the conclusion is quite simply that the planet's path is not a circle—it curves inward on both sides and outward again at opposite ends. Such a curve is called an oval. The orbit is not a circle, but an oval figure." The "oval figure" becomes the ellipse.

With this law, so much is dispensed with: the whole cumbersome machinery of Ptolemy and the epicycles of Copernicus. In their place, Kepler provides an elegant sweep of planets moving along the slight flattening and elongation of ellipses—sometimes barely discernible from circles—about the Sun. At the Buhl, no one mentioned what this represented in the history of science: namely, the triumph of raw data over a priori belief. How easy it would have been to ignore those few minutes of arc, to dismiss the numbers as somehow flawed, fogged over by an evening's revelry off the Danish coast. But Kepler worked with what he had, what had fallen into his hands in the tables of Brahe. And, as Hanson noted, this changed everything.

The *Astronomia Nova* contains, in addition, a statement of Kepler's second law. This proclaims an even deeper order in the heavens than the first. If you draw a line from the Sun to one of the planets—Venus, say—and you imagine the line sweeping out a region of the ellipse for an hour, as Venus moves, then Kepler's law says that the area being swept is always the same, hour after hour, regardless of where Venus is located in its long progression about the Sun. This is called the law of equal areas, and in general terms it states that "in any equal time interval, a line from a planet to the Sun will sweep out equal areas." There is, of course, no way that Kepler could have seen this, recorded it as fact. There are no ellipses, no lines or shaded areas in the sky. This is a pure creation, based only upon the tedious gaining of point after point and the fruitful workings of the mind.

It was seven years after he had departed Prague, in a time of personal turmoil, when his sister had died and his mother had been put on trial and condemned as a witch—a terrifying ordeal from which she was

68

finally released after 425 days—that he published his final work, *Harmonice Mundi* (1619). In this book he captured a fundamental unity in the motions of all the planets in a simple, memorable equation. Long after I had forgotten so much of what was taught at the Buhl on those humid summer mornings, I remembered the third law. It said, in effect, that if the period of a planet's revolution about the Sun were squared, and then divided by the mean radius of its path cubed, a constant would be obtained. For all the planets, the same number, as though a mysterious intelligence had willed it long ago. From tiny Mercury to Venus to Earth and outward to Mars, Jupiter, and Saturn, there was a plan for the heavens. And Kepler had seen it first.

How remarkable this law, when you think of it: Mercury with its high density and cratered surface, its fleet passage around the sun; and Venus, surrounded by a bright, toxic atmosphere and a thousand crowded volcanoes; Earth, endowed with an improbable ocean of liquid water, and a strange oxygen-bearing atmosphere; and Mars, absent now its long-departed rivers and seas, and its thin air in which oxygen is but a remnant, a trace; massive Jupiter and its storms that have raged hundreds of years; and Saturn, circled in the tiny particles of its rings. All of these, with all their differences, their disparate geology, physics, and chemistry, hew to the same unifying laws. And the planets discovered since then—Uranus, Neptune—along with the swift moons of our solar system, all so different and yet all, in that Keplerian way, the same.

The Dutch historian and philosopher E.K. Dijksterhuis tells us that Kepler's approach to science involved decoupling its practice and theory from theology and from long-embedded tradition. It advocated pushing science toward mathematical modes of thought and rigorous hypothesis testing and verification. Although these tenets are rarely made explicit, rarely written on sheets of paper or tablets of stone or elaborated on in our modern textbooks, everyone who does science today *feels* them as commandments. Every time I proposed how a lake had formed, or how a tiny particle of iron oxide carried with it, on its surface, a retinue of transition metals bonded in place, or how a molecule or atom of gas, in its parting of a sea of ions, moved or ceased in its turning, I

*Statue of Tycho Brahe and Johannes Kepler
on the hilltop overlooking Prague.*

felt them, and gave to Kepler, through the ages, a tip of the cap, a metaphoric bow.

Boltzmann must have felt those tenets, enough to have carved on his tombstone the equation that would quantify entropy for all time, for all those who, within and beyond the circling walls of Vienna, chose to inquire of that equation's truth.

❀

Toward evening, I climbed up past Prague Castle to where the monument stood. It is located on the site of Brahe's former observatory. The statue is cast in bronze and shows the two astronomers side by side. Brahe holds a large sextant. Kepler holds a scroll of data. He stands in a heroic pose, like one of the great explorers, Columbus or Scott. His gaze is fixed on the heavens.

I am not an astronomer. I doubt that I ever thought seriously that I would become one. But seated near the statue, in the shadow of Kafka's castle, I knew that the heavens were where it had all begun for me. Those nights when I was alone and looking into a vastness and a faint star-swirl, into the thin dust of what had been and what was to be, light that had come from a billion years ago, a past so large, so unimaginable, I could not contain my own smallness, when I felt, like I imagine Kepler had, the pressure of starlight on my hand and my face, and when in the darkness I could not tell where I ended and all of what was out there began, when there was no line of separation, and I could not even think the line, because it was all one thing.

The river, the red rooftops of Prague, the ancient bridge where in river-time the villages came and went, the Cathedral of Our Lady before Tyne, where the master with his gold-and-silver nose lies in stone; all of these in the rain-cleansed evening, along with the ellipses and laws of Kepler so simple in their absences, so bare-bones and pure, a direct line to Wolsthorpe and London and Cambridge, to Newton himself, as Hanson had said. In the heavens there is always solace, an evocation of dreams, a recognition of a place that holds us.

Florence

The Arno flows south out of snowy Mount Falterona toward the city of Arezzo, turns west then sharply north on its way to Florence. I came to this city because, more than any other, I associated it with the second of the great Copernicans, the astronomer and physicist Galileo Galilei. In Prague a week earlier, I had found that just being in a place—seeing its monuments and churches, its snug little streets and bridges—called up the revenants of memory and allowed me to see not only the history I had come to experience firsthand, but also, in a way, my own past and the fractured path that had led me to science.

Galileo had lived and worked throughout much of Italy—Padua and Venice, Florence, Pisa and Rome—but it was Florence that seemed to best capture the man and his life. When he came here from Padua in 1610, he was already famous for his lectures, scientific paper, and inventions, which included a thermometer. He had improved upon the design of the Dutch telescope, and at the age of forty-six, published his first major work, *The Starry Messenger*. Unlike earlier masterpieces in astronomy, *The Starry Messenger* met with immediate success.

I came to this little book of observations late in my life, when I happened to teach a liberal education course with my colleague, Barbara. Barb was a brilliant theoretical physicist, who, for fun and pleasure, translated *The Iliad* and spent her summers in the Greek Isles thinking about C-star algebra. We had chosen all of the books for the course except two, and we decided that we would each pick one of our favorites and add it to the list of readings. I chose Thomas Kuhn's *Struc-*

ture of Scientific Revolutions, a surprise best-seller on the history of the physical sciences. "They will love it," I confidently predicted. And she chose Galileo's book, which in my education as a chemist I had never heard of. "They will love it," she said. As it turned out, they hated Kuhn and loved Galileo, and through that single book, I discovered a scientist whose life and work I would later explore.

It is surprising that I had never heard of this slight volume. Growing up, I looked at the tiny moons of Jupiter and at Saturn's glorious rings in the shivering night through my neighbor, Newport's, reflector, the one he had built himself and even ground the lenses for in his own workshop. I had seen the moon magnified into gray craters and dust and put myself there, a lone figure walking awed toward the black horizon. In an early effort at lunar spectroscopy, I passed a beam of moonlight through a narrow slit in the blind that covered our attic window, and then through a prism I purchased from Edmund Scientific with money from my paper route. I am not sure what I was expecting to happen. Nothing did.

Like Gerard Manley Hopkins, I looked at night "at all the firefolk sitting in the air," and I marveled. I leaned precariously out of windows, perched on rooftops, felt the breezes of spring, tasted the sweet leafy air of October as I stared at the Pleiades and the Dipper and mused on those same timeless questions people have always asked about the stars.

❋

In Florence, I checked into the Hotel Morandi alla Crocetta, which was only a five-minute walk from the Piazza del Duomo and Giotto's Campanile, and just ten minutes from the Museum of the History of Science, where so many Galilean treasures are held.

In *Starry Messenger,* Galileo wrote of our nearest celestial neighbor, the moon. What he saw, for the first time in human history, was not the smooth circular disk that thousands of generations had seen before him, not the untarnished light of the night sky, made, as Aristotle had

Statue of Galilieo Galilei outside of the Uffizi Gallery in Florence.

said, of some ethereal fifth essence quite unlike our own upheaved and upheaving Earth. He saw an object that was rough and imperfect, whose crescent edges were jagged with specs of light, and whose surface seemed broken into "cavities and prominences."

Through long nights of careful observation, Galileo followed the movement of light and darkness across the face of the Moon and realized that he was witnessing the coming of morning to the peaks of mountains and the gradual illumination of deep valleys. Through his telescope, he was seeing the first landscape beyond the earth, seeing "more peaks shoot up as if sprouting now here, now there, lighting up within the shadowed portion." And he realized, too, something of the dimensions of the landscape he was witnessing, saying that "the grandeur, however, of such prominences and depressions in the Moon seems to surpass both in magnitude and extent the ruggedness of the Earth's surface"

Like Kepler, he made measurements. He did calculations, some of which were amazing. Not satisfied to know that there were mountains, he needed to know their height. With trigonometry and a set of distances obtained from his observations, he calculated that the lunar Appenines peaked at 20,000 feet and the Leibnitz Mountains at 26,000. His then-radical claim that the moon was "earthlike" was confirmed over the centuries by more powerful telescopes, and dramatically on July 20, 1969, when Neil Armstrong left his footprint as a lasting trace on the moon's windless surface.

Galileo was also overwhelmed by the number of stars that lay within the familiar constellations. He had intended to sketch Orion and to show the location of each of its stars, but he quickly realized that this would be a difficult task better left to another occasion. He wrote in the *Starry Messenger* that "there are more than five hundred new stars distributed among the old ones, within limits of one or two degrees of arc. Hence to the three stars in the Belt of Orion and the six in the Sword, which were previously known, I have added eighty adjacent stars discovered recently, preserving the intervals between them as exactly as I could." Among the six or seven bright stars of the Pleiades, he found scattered an additional forty, "no one of which is much more than half a degree away from the original six." The mysterious Milky Way, a faint dust of light to the naked eye, was revealed by the telescope to be "nothing but a congeries of innumerable stars grouped together in clusters. Upon whatever part of it the telescope is directed, a vast crowd of stars is immediately presented to view. Many of them are rather large and quite bright, while the number of smaller ones is quite beyond calculation."

The stars he was seeing in such profusion—and he was the first human being in the long hundred thousand year history of our species to behold such wonders—were tiny torch lights, tipped with flame.

They have, as he said, "the aspect of blazes whose rays vibrate about them and scintillate. . . ." These were so unlike the planets, which appeared as "globes perfectly round and definitely bounded, looking like little moons, spherical and flooded all over with light." And unlike the planets, the size and definition of the stars did not change with magnification. Only their brightness was enhanced by the telescope. This important difference convinced Galileo of the great distance of the stars and it explained one of the problems associated with the Copernican theory. Copernicus's Sun-centered universe had been criticized because it failed to explain why, if the Earth truly traveled in annual orbit around the Sun, there was no shift in the apparent position of a given star as viewed on an evening in January and an evening in June. Measurable parallax was expected. None was observed. But Galileo's observations suggested that stellar distances were truly vast and that the anticipated shift in position—the angle of parallax—was likely to be well below the limits of detection possible with seventeenth-century techniques.

In the hands of Galileo, a single instrument was beginning to erode the sturdy foundation of Aristotelian science. Copernicus's bold vision of a central sun and Kepler's geometry of the heavens, which replaced the Aristotelian circle with the elegant ellipse, were, by themselves, less than sufficient to challenge a system that had held firm for more than two thousand years. But Galileo's telescope offered something new. It was now possible to see that the distinction between the perishable sublunar realm and the constant eternal heavens was problematic. And if the ancient Greeks had had such an imperfect knowledge of the number and disposition of the stars, as Galileo had shown, what else had they missed?

Even more than moon and stars, there was the planet Jupiter, the crown jewel of the *Starry Messenger*. Anyone who has ever looked at Jupiter through a telescope or a set of binoculars from a darkened field has been struck by the presence of several small points of light in the planet's vicinity. Perhaps there are two or three, or, if fortune smiles, the entire visible complement of four.

The Copernican claim that the Earth moved about the Sun without losing its lone satellite, the Moon, was a serious conceptual problem not faced by geocentric theorists. While the observation of Jupiter did not provide a physical explanation for the association, it did show that a planet could travel through space with its moons in tow. And not just one moon, but four. Moreover, Jupiter was a kind of miniature solar system—a heliocentric model, for all to see, of satellites traveling serenely about a central body.

Could Galileo have imagined that his planet, Jupiter, and the satellites he had named for his patrons, the Medici, had fifty moons, and that the one he had seen, Ganymede, was still the largest in all the solar system and that Io was blazed with volcanism in its night skies? Callisto and Europa, beneath their crusts, moved with oceans of ice. There were rings he could not see with his small tube, and an atmosphere probed by a spacecraft named in his honor. The atmosphere was made of hydrogen and helium and racked with fearsome storms. A giant red spot, visible now from Earth, is itself a storm that has raged for centuries.

❦

Early the next morning, I hurried past the Uffizi Gallery, with its rooms of Caravaggios and Botticellis and Titians, and the usual large crowds bathed in the delicate dusty light of the city that conspired to produce the feeling that Henry James called Florence's "abiding felicity, the sense of saving sanity, of something sound and human. . . ." I was on my way to the Museum of the History of Science, not far from the Uffizi in actual distance, but remote in some conceptual way. The yellow Arno, with its bridges—the gentle parabolic arcs of the Ponte alla Carraia and the dark inverses that they cast in the water were the paths of thrown objects, of Galileo's projectiles.

Inside the museum, the stairways were empty and there was a lovely intimacy to the place, as though I had been invited to a private showing.

In the Galileo Room, the famed parchment telescope seemed like a slender reed, an object of art from a distant time, whose use, without

Bertini fresco of Galileo demonstrating his telescope to the Doge of Venice.

historical record, might only have been guessed at. It was a wooden tube covered with paper and equipped with a double-convex lens and a plano-convex eyepiece capable of fourteen-fold magnification. Little more than a spy glass, this simple tube could carry you to the peace of other worlds. No wonder that the curator of this museum had, in one of the angry periodic floods of the Arno, risked his life in pursuit of this very telescope.

The same room also holds a curious relic. It is Galileo's middle finger, positioned prominently in its own little monument of marble and glass. The finger had been detached from Galileo's hand in 1737 when his remains were transferred from a closet near the Chapel of Cosmos and Damian to its present location at the mausoleum in Santa Croce. I stared at this finger, thinking that its placement among the scientific instruments was quite bizarre, until I read the reverential inscription that had been written by Tommaso Perelli, the eighteenth-century professor of astronomy at Pisa who had been responsible for a Galilean revival of sorts: "this is the finger which covered the heavens and indicated their numerous space. It pointed to new stars with the marvelous instrument made of glass and revealed them to the senses."

※

Walking through the streets of Florence that evening I imagined Galileo strolling past the same piazzas, along the River Arno, with its ancient bridges looking up toward the Tuscan Hills. I returned indoors to look for a while at Michelangelo's *David* and wondered at the mysterious process by which marble—formless, earth-compressed calcite, the remains of microscopic coccolithophores—could be turned by the human hand into this perfection.

Earlier I had seen the statues that Michelangelo had intended for the tomb of Julius II. In these massive blocks, you could see bodies coming into being. The faces, arms, and legs were not defined. These were meant as allegories of the soul imprisoned in the body. But in the context of my travels, I took them to be a metaphor in stone, a metaphor of an idea emerging.

In 1610, in Prague, the Tuscan Ambassador handed a slim copy of the *Starry Messenger* to Johannes Kepler. In a short time, the appreciative Kepler responded with a pamphlet extolling the work of Galileo. Sixty-seven years had passed since the death of Copernicus; fifty-seven years had passed since Giordano Bruno was burned at the stake for his heretical beliefs in a sun-centered universe. But in 1610, there were at least two powerful figures in Europe yearning to bring forth a new system.

※

We are rotating on an axis at about 1,000 miles per hour, and revolving about the Sun even faster, at 67,000 miles per hour. We rarely think about these things. We never experience them, although these are tremendous speeds. Galileo recognized that this was a serious problem for the ideas of Copernicus. He needed a physics for a moving Earth. He needed to explain why an arrow shot straight upward would fall nearby and not well behind as the Earth turned swiftly east beneath it.

Much of Galileo's physics can be found in his *Dialogues Concerning Two New Sciences*. The dialogues, composed when he was blind and under house arrest, were published in 1638, after his famous trial in Rome and after the death of his beloved daughter. The book appeared in English in 1665 but was consumed by the London fire the following year. It was not published in English again until 1730. The dialogues are anything but arcane. They are clear and modern, and they lay out the foundations of kinematics: the acceleration of falling bodies, the movement of projectiles along a parabolic path, the nature of inertia, the principle of superposition.

But it was the principle of inertia that seemed to answer the early questions about the Earth's turning. If an arrow were shot upward, it would have two kinds of motion: it would move in response to the bow, but it would also partake of the Earth's eastward rotation. In effect, it would be hurled to the east with the same frightening velocity as the turning Earth itself. The same would apply to every thrown rock and baseball, to every bird on a limb that ever chanced to dive for the worm below.

There is a more or less famous demonstration of this captured on film. It involves two black-and-white photographs by Bernice Abott. In one photograph, an old model train stands on a track. From its smokestack, a tiny ball has been fired, and the ball has been photographed using a stroboscope. The white ball shoots upward from the *motionless* train against a black background, reaches its zenith, and returns to the small hole from which it was fired. Nothing dramatic here. But in the adjacent photograph, the train is *moving*. It is moving at a constant speed along a smooth track. The spring-loaded ball is again projected upward from the stack and the camera captures its rise and fall. Where will the ball land? Behind the moving train? Back in the smokestack?

If you're not familiar with the principle of inertia, the result may surprise you. In the photograph of the ball shot from the moving train, the camera records points of white arranged in a parabolic arc. Each point lies just above the smokestack of the train, which is racing along below. In the end, the ball lands right back in the hole from which it was ejected, not behind the train.

This kind of motion—projectile motion—was first explained in the *Dialogues*. It requires that the ball has not only the upward thrust imparted by the spring, but also the motion imparted to it by the moving train. Like the arrow in flight, the ball in flight partakes of the motion of the body from which it was released. I have seen this experiment carried out from convertibles and from speeding bicycles with the same results (if you are generous enough to subtract the wind).

Dialogues was published in liberal, Protestant Holland in 1638. Galileo died four years later, the year of Isaac Newton's birth.

✧

Beyond the piazza, off the Via Della Vigna Nuova, lay water. As I drew closer, I could hear it, flowing over a shallow spillway. It drowned out everything—the sound of the buses and tourists, the annoying Vespas, and though I could see debris from the night floating downstream, the Arno seemed a very old and pure thing.

By eleven o'clock, I had climbed high above the city to the Piazzale Michelangelo. There was a café there with a view of the Tuscan hills, where I drank a bottle of Tuborg and ate salted peanuts and looked out over the city of Galileo. So much of what he had done came back to me: the studies of motion, which explained how objects (the baseballs and footballs of my youth) could move the way they did on a moving Earth; the trial that had led him to recant his beliefs in the visions of Copernicus; the brilliant physics contained in the *Dialogues Concerning Two New Sciences*; the inclined plane, with its five bells clamped to the frame, and the water clock he had made to time the silver ball as it rolled. The flowers all around were red and pink and picked up the hues of the slate rooftops of Florence so that the whole scene was vibrant with color and depth and sunlight across the whole of the valley.

London
and Jupiter

Kids I knew, from the neighborhood, from school, from some team or other, had grown up to do very different things with their lives. Jerry had become a priest. So had my brother Eugene. McMullen had become a builder of skyscrapers. The two of us had climbed to the top of one of them on a cold night in Chicago, when the building was still an open skeleton of girders looking out into nothingness, and for a moment I thought I was in a Saul Bellow novel, standing thirty-five stories above cold Lake Michigan. Luke, a decorated colonel, had been to Vietnam; McDowell, "Sudden Sam," as we called him for his blazing fast ball, became a famous pitcher for the Cleveland Indians and set records on the mound you could only dream of; Smitty became a draftsman; my sister, a budget director for a large university; and my good friend Larry became an accountant and owned a house on the water near Daytona Beach.

In my thirties, I taught at a college in Ohio. I had a laboratory which, by today's standards—or, perhaps, by any day's standards—was less than humble, equipped as it was for little more than a few chemical analyses. I recall passing by a faculty office one day. Elizabeth, a professor of literature and the arts, often posted quotations that I found thought-provoking. This time there was a line by Albert Camus, whose novels I had read in an ambitious undergraduate course on tragedy, and whose works I found more than a little disturbing. This quote, it turned out, was from his youthful writings, and I immediately copied it onto a crumpled sheet of paper that I put in my wallet as if someday I might actually need it. Camus wrote: "A man's work is nothing but the slow trek to rediscover through the detours of art those two or three great and simple images in whose presence his heart first opened." The words arc you back into the past, into some sort of personal sediment where the fragile, indecipherable shards of your early life lie, and where, on your knees, brush in hand, you had to sort and sift for the smallest clue if you ever hoped to discover those two or three great and simple images Camus spoke of.

My mother's interests were Latin, which she taught with passion, and grammar, which she preached to us every day. My father had gotten through high school, but he was a practical man, a civic leader. In our sports-obsessed neighborhood, where there were only three seasons—baseball, football, and basketball—I can't recall Newton ever being mentioned.

Perhaps I had heard of him for a few minutes in world history. I think he merited a line or two in a chapter on Pope Urban VIII and Innocent XIII. It wasn't until my senior year, when I took my first course in physics, that I actually studied his laws and achievements. These were offered in the abstract language of high school textbooks, and as usual, I failed to see that there was a human being behind the science. These laws just seemed to have dropped out of some "sciency" mist of long ago, and were now the stale truths of memorization from which a thousand little problem sets would be constructed.

My first Newtonian experience came, however, when I was fifteen or so. I was caught up in the excitement generated by *Sputnik* and the achievements of our space program. For months I had been looking heavenward, dreaming of stars and planets and what it would be like to visit them someday. With all the writing and talk

Newton as depicted by William Blake.

about space at that time, with *Popular Mechanics* and Scott Cross-field and his X-5, and Edwards Field smooth on the shimmering salt flats of California, it seemed as though it would be only a decade or two before one of us, perhaps even myself, would be standing on the dust plains of Mars looking over landscapes of untold beauty and doing scientific and heroic deeds beyond the dreams of most mortals. I had read Ray Bradbury's *Martian Chronicles*, and it seemed as though launching a rocket to a nearby planet would one day be as simple as catching a trolley for downtown Pittsburgh. Of course, Mars, with its great system of canals and its hostile inhabitants with their hovering saucers and death rays, had no equal in the solar system for pure adventure.

So my friend McMullen and I—like members of some cryptic medieval sect and ignorant of all things Newtonian or even vaguely scientific—gathered the materials we would need for our budget pro-

totype spacecraft: the aluminum tubes, the plastic pencil sharpener
that made a great nose cone, some aluminum sheets we could cut into
fins in McMullen's workshop. Soon we had our shiny miniature of
brushed metal nearly ready for its epic voyage into the unknown. All
we needed now was the fuel. How I did it, I'm not sure, but I convinced my father to purchase the necessary chemicals from the Fisher
Scientific Company, purveyor of all things explosive and wonderful.
"It's for science!" I said.

In a week we had the potassium nitrate, the charcoal, and the sulfur. Like alchemists, we played around with the percentages, mostly by
saying, "Okay, this looks right," and then setting it on fire. There was
nothing quantitative about this. We had no scales and had to judge
everything by eye. Eventually, through rough experimentation, we concluded that we needed mounds of potassium nitrate, a bit of charcoal,
and a smidgeon of sulfur to get the desired effect. I still have no idea
how well we approximated the 75/15/10 ratios used in gunpowder, but
I suspect we were close.

Our first launch, preceded by lots of chatter and sweat and completely aimless motion, was more or less typical. On the pavement,
several feet from the back porch, in full view of the neighbors, our
rocket stood on its wooden pad, a proud ungyroscopic cylinder aimed
into space. A hole had been drilled through the top of the launch platform to allow the exhaust free passage, and we expected a long luminous yellow flame to emerge that would form a sharp tip at the end.
And so we made allowances by making the pad five feet high so that
the nose cone rose just above the hedges and the rocket stood high
above the summer grass. This launch and all around it would be a
thing of beauty.

And indeed it was—for perhaps a microsecond or two. As I recall,
the rocket shuddered a bit at first, wobbled into the air, and then, to my
surprise, rose, curved toward the house and landed on the roof. There it
lay like a hissing railroad flare threatening to burn the place down. I ran
into the kitchen, through the hallway, up the stairs, into the bathroom,
out the window, and onto the roof. I gave the furiously burning tube,

sadly no longer a rocket, a good kick that sent it off and down and into the bushes, where it burned, crackled, and singed until it was out.

Our subsequent attempts were improvements, but only in the sense that our larger, more stable, and more powerful rockets inflicted more impressive damage on the surroundings. Our next missile, which did take off in the right direction, suddenly boomeranged back toward the porch, blew a huge smoking hole in my mother's new screen door, shattered our lovely red nose cone to smithereens, and lay burning by the wooden back door where it made a deep black scar before I could retrieve it. My patient mother, who had been watching all of this from behind a window, finally came to the door and suggested, the advancement of science notwithstanding, that we consider abandoning this quest for Mars.

On the last try, our spacecraft actually lifted off. It rose straight up, possibly a hundred feet or more, against the moon. For a moment, there was utter perfection—the glowing rocket, the night sky, the smell of burnt sulfur blended with the moist earth.

But then came the unexpected plunge, straight down, straight toward the doghouse that our nextdoor neighbors, the Reardons, had erected for Pepper, the black-and-white dog whom they had recently rescued from the pound. Pepper was in his house napping. Our rocket, as if guided by the minds and equations of some brilliant, nefarious engineer, smashed directly into his roof. Poor Pepper. Never had a dog moved with such speed or with so many high-pitched whelps. Never had I seen the Reardons come flying out of their house with such urgency—mother, father, my friend Raymond—to be followed by my own family's appearance on our back porch. At this point McMullen and I just stood there looking at one another, silent as the night had been earlier, knowing it was all over. Rocket summer. Voyage into space. The whole ballgame. Soon I was back to reading Werner von Braun, Ray Bradbury, and *Popular Mechanics* and—best for all concerned—living the rocket life only in my imagination.

When, years later, we finally got to our high school physics course and listened to the first lecture by Sister Amy on Newton and rocketry and how physics mattered in everyday life, McMullen and I glanced at each other and I felt I knew what he was thinking: "So *that's* what we were missing back then!" From that point on, I wanted to know more about this science. I wanted to know what else in my youthful experience—the curve of a fly ball, the path of the moon, the pressures and forces and thrills of the amusement park—this man and his physics had explained. His world, after all, was mine.

※

Many years later, in the midst of my historical sojourn, I was in the land of Newton's life and work. The Newton Memorial at Westminster in London has a calmness and dignity about it that contrasts with the commotion outside, the flowing crowds, the familiar black taxis hurtling by. The figure of Newton reclines, like Caesar, in loosely fitted robes beneath a globe of the Earth, his right arm draped across a collection of his greatest works: *Principia,* of course, and the *Optics.* The instruments of his research, telescope and prism, stand in bas-relief. The inscription, in Latin, which tells of his achievements in mathematics, astronomy, and optics, ends with the words: "Mortals rejoice that there has existed such and so great an ornament of the human race"—words which struck me as a proud proclamation of ownership, not merely by England or by any single country or race, but by the whole of a grateful humanity.

Unlike Boltzmann, Newton was not born to middle class privilege, with private tutors, surrounded by a city filled with music and philosphy and at the very height of its influence.

An eighteenth-century painting by J.C. Barrow shows the farmhouse in tiny Woolsthorpe where Newton lived as a child. It appears as though it might have been dropped on the landscape by a tornado, like something out of the *Wizard of Oz.* Slightly crooked, it stands in furrowed fields, its tall chimneys aimed at a gray English sky, the small gate not quite plumb, with slender trees on the dusty horizon.

The stillness of Barrow's scene belies the turmoil of Newton's youth and times: the English Civil War, in which his father had died before Newton's birth, in 1642; the marriage and departure of his beloved mother, Hannah, from the family home and her subsequent return two years later; the school in Grantham, where he was sent at twelve and where, among his classmates, he was called "stupid" and "sissy." Near the end of his life, Sir Isaac Newton, the most famous man in all of eighteenth-century Europe, would take pride in recounting to his French biographer that, after long provocation, he had finally confronted one of his schoolyard tormentersy, and had dragged him off by the ears and beaten his head against a wall.

But to offset this, there were the hours of boyhood contemplation, the life of the mind discovered as a kind of balm against the terrible realities of the world. The Clarkes' home at Grantham, where he boarded, had a library high in the house, up through narrow doors into an attic. It contained the writings of Descartes, Christian Huygens, Kepler, with their odd tangle of science and mysticism. In the Clarkes' room he found peace, a conduit to places far beyond the English countryside. What thoughts must the beautiful ellipses of Kepler have evoked for him, drawing his mind, as they had drawn Kepler's, into the deep silences and geometries of space?

When I left Westminster, I walked to a nearby park, into the sunny greenness and the noontime crowds. I was struck by the movement and hum of things. Newton's ideas had really gotten to me. It was like going out into the street after you had seen a great film or finished a great novel. Everything was somehow more vivid and alive. On the walkway, a woman pushed a carriage. A rosebud dipped in the breeze. The lake water rippled and waved, a leaf adrift on its surface. A pelican stretched its wings as if to fly but then folded them back into itself. Every blade of grass bent and straightened. The sounds of London were the sounds of motion, too, the cars and taxis and buses. Above me, there was a plane flying across a nearly invisible trace of quarter moon. It all formed a tableau of motion like a Calder mobile.

Motion was at the very heart of Newton's work, and so was the idea of force. For an object to move or to change its path, a force must be applied. Imagine: the woman pushing the carriage along the clay path lets it slip from her grasp. The carriage continues to move in a straight line, and with no resistance from Earth or wind or friction of any kind its path extends straight to the edge of the world. It rolls on and on at a constant speed, forever. Only a force could change that. In essence, the imagined carriage is obeying the first law of the *Principia,* Newton's great work on dynamics. In formal language: "Every body perseveres in its state of being at rest or of moving uniformly straight forward, except insofar as it is compelled to change its state by forces impressed upon it."

But the world is shot through with so many pushes and pulls. The rose I am watching is utterly still. Then, all of a sudden, it dips toward the greenish lake. The wind exerts a force, and the movement, from stillness to some velocity, is an acceleration. The force has caused the rose to accelerate. In the *Principia,* in his second law of motion, Newton expressed it this way: "The acceleration produced by a particular force acting on a body is directly proportional to the magnitude of the force and inversely proportional to the mass of the body." In every physics book, though not in *Principia,* the law appears in its familiar form as $a=F/m$; or more commonly as $F=ma$. In a strong and forceful wind, the rose, with its tiny mass, is almost jerked from the branch, while the massive pelican on the lake shore scarcely moves at all.

Newton's understanding of force extended beyond the confines of a single park, a single planet. Two objects, already mythical, illustrate this. They are the apple and the Moon. No two objects could be more different: the apple is a thing of this earth, come out of its soil, hard and edible and smooth, small and close at hand. Galileo could have described how it fell, had he dropped it out of the white tower at Pisa, accelerating toward the green lawn below. The Moon, on the other hand, is distant, an object of the heavens, a full brightness over the cornfields of Indiana and Ohio. Yet, for Newton, the fundamental facts of these objects lay not in their differences, but in their sameness; how in their apparent diversity they are unified parts of the same fabric of nature.

What Newton recognized was that both apple and Moon were drawn to the Earth in a certain way, accelerated by a certain force that extended not just to the top of the highest apple tree but all the way to the Moon and beyond. And he reasoned that the force that pulled the apple downward also tugged at the Moon. Like the apple, the Moon was falling and falling, but its speed, its orbital velocity, prevented it from ever actually reaching the Earth. The force was the mysterious force of gravity, operating always according to law, to the law that Newton himself wrote and that we know as the law of universal gravitation. This powerful generalization, applying equally to suns and moons, apples and chestnuts, says that the force of gravity that objects exert on each other depends directly on their masses and diminishes by the square of the distance separating them. This makes it possible to calculate forces and, when joined with the other laws of the *Principia,* to explain the motion of the apple in the garden and the Earth-encircling Moon. In fact, it makes it possible to launch a space shuttle and—my childhood dream—to land on Mars.

Newton had said that he stood on the shoulders of giants; among these, first and foremost, were Kepler and Galileo. Newton used his abstractions of a universal gravitation and a set of forces that depend upon mass and acceleration to derive both the laws of Kepler and the insight of Galileo's that the velocity of falling bodies is independent of mass. The provable laws of earth and sky seemed to tumble magically from the abstractions of mind, to be illumined in all the monastic rooms by it: with Newton, the revolution begun by Copernicus was 1543 is realized. Modern science was born. And with it, the world we see and touch and hear, its daily marvels and wonders beyond counting. It was true, as Alexander Pope had written in that couplet he meant the stonecutters to inscribe at Westminster: "God said Let Newton be / And all was light."

❧

Dayton is not far from where I live. Sometimes it is not easy to think of it and Newton together. Out of the cornfields, in a vast acreage you do not quite expect, there is the sprawling National Museum of the United States Air Force. When you enter it and see the statue of Icarus in the foyer and you take a map of the place from a uniformed vet, you wonder how you are ever going to see it all. The groups you pass in the broad hallways are speaking French or German or Japanese and it's as though all the world has come to celebrate this new thing in the history of our species. A diorama of the Wrights' bicycle shop, with its chains, workbenches, and drawings, and the brothers themselves bent to their tasks, takes you back to the clean, slow-moving days of 1903, when the spindly plane lifted and stayed aloft.

Beyond the Hercs and Hueys and Starlifters that I knew from Antarctica, and the plane with the ball turret gunner, whose death had been immortalized in Randall Jarrell's 1945 poem ("When I died they washed me out of the turret with a hose"), there is the silolike structure that housed the rockets, the *Titan I* and *II* and the *Jupiter*. Next to me, someone said, staring straight up at them, "God, there are a lot of smart people in this world." The plaque that explained these reaction devices recalled Newton's third law: "In a rocket, the *action* is the expulsion of gases from a nozzle at the open end of the rocket body. The *reaction* is the movement of the rocket body in a direction opposite that of the expelled gases." Although rockets had whizzed through the air in thirteenth-century China and had found their use in countless displays ever since, Newton's formulation brought rocketry into the fold of science and number, into these rooms of exploration where Moon rocks and Moon dust lay.

❈

Toward the end of the last century, I spent two years teaching at a college in south Florida. It was located in the quiet town of Jupiter before much of the fevered land rush had begun, and Jupiter still had that rustic, swampy, heat-burdened feel of the old state. You could be out among giant ferns and alligators and acres of sugarcane in minutes. In the morning, Wanda and I liked to walk on the nearby beaches, and sometimes it would occur to me that this, too, was a Newtonian province—a place with tidal rises and falls, whose fishermen, with their long poles buried in sand, were figures in a world that the predictable low tide had created. It was a world that followed directly from the law of universal gravitation, which was created in the beginning by the forces of Moon and Sun on the yielding fluid of the sea.

One night in Jupiter, we were standing on the grassy shores of a clear pond near the condo we were renting. A small crowd had gathered and there was some genial, neighborly conversation and the soft flutter of anticipation. People were facing north, turned into the flatness of the Florida landscape, toward the horizon. Then it came, without sound: a long trace of whiteness against the fading pink of the evening sky. There was a full yellow moon, large enough that you could almost make out its Galilean features, and for a moment nothing existed but the Moon and the great trail of whiteness, the rocket from Canaveral and its attached shuttle, ascending.

Bern and Washington, D.C.

In Washington the cherry blossoms came early. A young new president took the oath of office after a historic election, and the poet Robert Frost read at his inaugural. I attended a meeting of the American Chemical Society and listened to the century's greatest chemist, Linus Pauling, deliver a lecture on electron-deficient compounds. I was away from home for the first time, dazzled by the city's monuments, by the words of Lincoln carved in stone, by a sense of something large and powerful. Just walking the neatly laid grid of streets Pierre L'Enfant had designed, I felt I had arrived at the center of the world. Even the campus of Catholic University, with its byzantine shrine that I could see from my window, impressed me.

Among the few possessions I brought was a copy of Paul Arthur Schilpp's collection of essays: *Albert Einstein: Philosopher-Scientist*. I placed it prominently on the bookshelf in my dorm room. I had read only parts of the first chapter, written by Einstein himself, so the book was little more than a decorative item.

When Bill Polking, a law student and the dorm RA, came to the room to introduce himself, the book was the first thing he noticed. "You must really be serious about science," he said, paging through an assortment of critical essays by Percival Bridgman, Louis de Broglie, Max Born, and other luminaries I had never read and wouldn't have understood then if I had. Still, with Polking, the book started a conversation and a friendship, and provided me with an identity that I was not uncomfortable with.

"So what do your parents do?" Polking asked, probably expecting that I was the scion of theorists going back generations. "Well," I said, "my dad works with the railway company—trolleys, buses, scheduling, that kind of thing. And my mother teaches Latin. She has an MA," I said, with a note of pride that he easily detected. Polking smiled at that and, handing me the volume on Einstein, said, "I think her son will have his PhD some day."

I was a freshman, uncertain of everything, lost in a great city. A frightened eighteen-year-old who had come to college with Schilpp's book on Einstein as a defense. My horizons hardly extended beyond the course catalog, the baseball field, the dining halls of the college, and the few friends I had made on campus and at the mixers at nearby Trinity. A PhD? I had never thought of it. But from that moment on, I would. And Polking's words, casually tossed off as he leaned against the unmade bunk beds of our tiny room, became a beacon.

❧

Before I visited Einstein's apartment in Bern, I climbed the steep hill above the town. A spring breeze moved the branches of the trees. Down in the valley of the Aare, the steeples of the churches were green-weathered copper and the rooftops of the city were red. The river ran swift with clay from the peaks and snowfields of the Alps.

Bern was unexpected: small, ancient, provincial, its massive sandstone buildings like fortresses, its narrow streets and arcades from another age. The trolleys ran silent, smooth, and on time—a system my father would have readily understood and admired. Tall statues of armored bears stood with their colorful shields and swords on small islands in the middle of the streets, and there were flowers in the boxes of every window.

Not long after I arrived, I went to the apartment at 49 Kramgasse. On the first floor there was a bar. On the wall, a large poster read: ALBERT

The clock in Bern's Old Town, near Einstein's apartment.

EINSTEIN'S BERNER JAHRE: 1902-1909. I sat down, ordered a beer, and pulled out a copy of *Albert Einstein: Creator and Rebel* by Banesh Hoffmann, which I had brought along. It was a book that Barbara and I had used in our courses and that she had considered one of the best portraits of Einstein's creativity and deep humanity. When I first read it, it evoked what was large and distant and mysterious. The whole shape and structure of the universe was laid out in its pages, as were the life and vision of the man who had seen far beyond Isaac Newton.

Einstein was born in Ulm in 1879. A year after his birth, his family moved to Munich, the Bavarian capital, where his father set up a small electrical factory. Banesh Hoffmann, whose book I had before me, tells how at age five the young boy was given a compass by his father. He held it in his small hands, marveling at the movement of the silver needle as it swung north. Always the same, no matter how he shook it. What invisible power was at work here? The God of utter strangeness moving in the most commonplace of things. Hoffmann tells us that this simple object, as common as a grocery scale or a falling apple, awakened in Einstein a curiosity that would remain throughout his life and would cause him to question what others often took on faith or authority. It was in Munich, too, that he discovered a love of mathematics, especially Euclid's "holy geometry," with its clarity and certitude.

Eventually he enrolled at the Polytechnic Institute in Zurich and graduated from there in 1900. Appointments to various teaching posts, all temporary, followed, until he was offered a position here at the patent office. To me it always seemed that being a clerk and a civil servant in Bern was not the best point from which to launch one of the great careers in the history of human thought, but Einstein would later claim that the patent office had afforded him the time and economic freedom to pursue his studies of fundamental physics.

After I had finished my drink, I climbed the stairs to Einstein's apartment. He had lived there during the great years with his wife Mileva. There is a soft, dreamy quality to the place. The wood appears gilded. The curtains fall nearly to the floor. There is a fireplace, near which I imagine him seated, staring at the sparks as though they were tiny stars being tossed into the blackness of space. On the wall, there is a photograph of the young clerk. He is dressed in a suit and tie, standing behind a tripod writing table. He is young and his hair is black. On the wall, some of his more famous words are written: "Blind belief in authority is the greatest enemy of truth." "Discovery is not the result of logical thought, even though the end result is intimately bound to the rules of logic." And, of course, there is the great equation, framed with the date, 1905, written above it: $E = mc^2$.

On the wall, there is a list of the papers that he wrote in this very flat and published in 1905—a year that is often referred to by physicists as the miracle year, the *wunderjahre*. He was twenty-six. Included are the famous works on Brownian motion, light quanta, special relativity, mass-energy equivalence, and the dissertation on molecular dimensions. Any one of these would have made for an astonishing career. There are photographs of Bohr, Newton, the three members of his Olympia Academy, and the young, dark-haired Mileva Marić, his first wife. "Those days in Bern were indeed happy days," he wrote.

Einstein's flat on the second floor at Kramgasse 49, Bern, Switzerland.

If you brush aside the long white curtains, ornately embroidered at the bottom, and look out onto the cobbled streets of the city, you can see the old medieval clock. It is a hundred yards away from Einstein's window. It is a thing of great beauty and dominates everything in sight. Perfect for the man who would forever revise our view of time.

❀

The first time I took chemistry, I was a junior in high school and, since chemistry was offered only once every two years, half of the students were seniors. Almost from the beginning, I felt that I understood this subject. Others seemed to be struggling, but I was not. The mole concept was perplexing to my classmates, this strange bridge between the invisible world of atoms and the world of real, weighable objects that we occupy every day. The periodic table seemed to most just another of those onerous exercises in memorization rather than the near-miraculous system of matter you could put on a piece of plastic the size of a credit card and carry around in your wallet if you chose. Even the Bohr atom, with its K, L, and M shells of electrons, like so many rings of caramel wrapped around a sugary center, was viewed as an improbable construction to be mentally stored for a week and then forgotten for a lifetime.

But much to my surprise, given the advance notices, I liked this stuff. And even more surprising, I was actually good at it. Sister Amy, who taught the course, said, "Have you ever thought of being a chemist?"

Back at the Hotel Glocke, in the heart of the old town, just a few minutes and many decades from Einstein's "Street of Time," I sat down in the empty dining room, with its heavy wood paneling and quaint checkered tablecloths to do some writing about Einstein. The accommodating owner agreed, even though it wasn't dinne time, so I spent a few hours making notes and trying to connect some random thoughts.

One image that returned to me from that high school chemistry class was of an observation we made through a simple microscope. The experiment had been set up for us, and all we were asked to do was look through a lens at what appeared to be a drop of water. "I want you to focus on a single particle," Sister Amy said. "Pick out oné and write down what you are seeing." My partner for this was McMullen, as usual, and the two of us were able to compare notes and prepare a single sheet of observations.

What I saw was motion. The particle, once I adjusted my eyes, was moving. But it was not moving in any clear direction. First it would go right, then left, then take a short upward hop, and then left again and then down and down again and out of the field of vision, then back again. The motion was erratic. It was random. I could not see order in it at all. "Look at this," I said to McMullen, with a note of excitement in my voice. "This is strange."

McMullen, who was six-four and skinny and wore wire glasses and thought mostly about basketball in those days, looked over at me with a frown, a little bored, a little skeptical about my enthusiasm. He removed his glasses, peered down into the lens, and kept his eye to the glass for a long time. When he came back up he said, "Interesting, Green. Never seen anything like it. What's going on? Atoms, I'll bet. It's got something to do with atoms." Then he smiled and said, "This is a chemistry course, isn't it."

We had just seen something that had been discovered more than a hundred years before by Robert Brown, a Scottish botanist, in 1827. Brown observed this erratic motion of microscopic particles initially with pollen grains, so he thought it might be a manifestation of some kind of life force. But on further experimentation, he saw the same kind of zigzag behavior with dust and other inorganic substances, and so concluded that it was a general phenomenon.

An explanation of Brownian motion was given in 1902 by the Swedish chemist Theodor Svedberg. Svedberg's solution to the problem was based on the earlier conviction of Maxwell and Boltzmann that matter is composed of atoms and molecules in constant random motion. In a liquid such as water, a microscopic pollen grain would be under constant bombardment by the surrounding molecules. But given the movement of the molecules, it seemed likely that the number of impacts would not be the same on all sides of the grain. At one instant, more molecules would be striking the left side and the particle would dart to the right. And at other times, more would be impinging on the left side and the grain would move to the right. This unequal bombardment led, according to Svedberg, to an aimless, irregular movement, like the walk of a drunken sailor about a lamppost.

This was purely qualitative. There were no numbers, no calculations, no predictions. But all of this changed in the city of Bern, a few blocks from where I was writing. In his dissertation, Einstein had developed a molecular theory of liquids based on the statistical concepts of Maxwell and Boltzmann. They had thought of the molecules of a gas as being in constant motion, with velocities and energies distributed over a broad range of values. The temperature of a gas was related to the average molecular velocity. Using their purely theoretical model, it was possible to derive easily testable laws, like those of Jacques Charles and Robert Boyle.

Einstein applied this statistical approach to liquids, and in a separate paper, published in 1905, he devised a test for his model. Suppose a tiny particle, say a grain of pollen, were placed in water. If we could see it, what would its motion look like? He performed the calculations, treating the water molecules as tiny Newtonian masses and using a statistical approach. From this model, in which matter was viewed as a jostling substratum beneath its apparent surface quietude, he predicted that the pollen grain would move in random fits and starts. He had never heard of Brown or Brownian motion. In his autobiographical sketch, which appears in the volume I carried with me to Washington, he writes:

> My major aim in this was to find facts which would guarantee as much as possible the existence of atoms of definite finite size. In the midst of this I discovered that, according to atomistic theory, there would have to be a movement of suspended microscopic particles open to observation, without knowing that observations concerning the Brownian motion were already long familiar.

Boltzmann had never read this paper. He was off in California in 1905, dining with the Hearsts, taking the train through Yosemite, complaining to the porters that there was no wine through great dry stretches of the American heartland. Einstein, using Boltzmann's statistical approach, and assuming the existence of molecules, had found the way to explain Brown's curious observation. In the end, even Mach and Ostwald would have to concur.

What if Boltzmann had known? What if he had been in correspondence with the young patent clerk with whom he had so much in common? Would that day in Duino have turned out differently?

Boltzmann's Tomb

❈

When Barbara and I were teaching together, we always included a lecture and a discussion on another of Einstein's signature works from this city, the paper on the photoelectric effect. In her lecture, she included the fact that it was this paper that had won him the Nobel Prize, and not his more famous contributions to relativity. This was always news to our students. My own seminars included a reference to the unpopular Thomas Kuhn, some of whose ideas about scientific revolutions and "paradigm shifts" applied here. Kuhn had in some ways long been a favorite of mine, not because I thought he was right about the nature and history of science (he wasn't), but because he had started a fruitful dialogue that seemed to engender far more passion than anything any academic philosopher could ever have dreamed of. Even scientists, who were rarely interested in what philosophers had to say about them, had heard of Kuhn and his masterpiece, *The Structure of Scientific Revolutions*, and they argued about his views. When I studied him for the first time in Pittsburgh, the professor, Larry Laudan, who was usually a model of cool restraint, could not help fulminating over Kuhn's "core flaws" and "hopeless positions." I have not seen the now-legendary Larry Laudan in years, but his writings suggest that in all his wanderings from Pittsburgh to Virginia to Honolulu to the National Autonomous University of Mexico, his views on Thomas Kuhn have not changed much. I like to think of him as a kind of philosophical, but sober, Malcolm Lowery, seated at a small café beneath Popocatepetl, sipping tequila from the narrow rim of a *caballito*.

The beer at the Hotel Glocke was unusually good, so I kept drinking it well into the evening. It seemed a natural accompaniment to the photoelectric effect—at least in my mind. The photoelectric effect, before Einstein came to Bern, was a great mystery. It had been known for some time that if you shone a beam of light on a (negatively charged) metallic cathode, say a sheet of zinc, it as possible, under the right conditions, to generate an electric current. J.J. Thomson had shown by

1898 that the electric current consisted of a stream of electrons moving from cathode to anode. And by 1902 Philipp Lenard, one of Boltzmann's favorite colleagues in Vienna, had demonstrated that these electrons, regardless of what metal they had come from, all had exactly the same properties.

The expectation, based on the dominant theory of the day, which was the wave theory of electromagnetic radiation as developed largely by Maxwell, was that a more intense beam of light would liberate more electrons from the metal surface and thus generate a stronger current. But, in general, this was not the case. You could illuminate a zinc cathode with red light for as long as you wanted and nothing would happen. No current at all. Or you could illuminate it with blue light, which has a shorter wavelength and a higher-frequency, and still nothing would happen. But as soon as you turned on an ultraviolet lamp, which emits a shorter wavelength and higher frequency (and hence more energetic) radiation, and directed it at the zinc cathode, a current would begin to flow.

What was going on? There seemed to be a threshold frequency below which no electrons could be ejected from the metal surface. The idea, propounded by the wave theory, that eventually even low-frequency radiation would deliver enough energy to the metal surface to kick out electrons, appeared not to be true. In the language of Thomas Kuhn, the wave theory of light had been confronted by an anomaly.

In the quiet of the patent office, in the small spaces of the apartment on the Street of Time (as I came to think of it), Einstein had another idea. Maybe, in the case of the photoelectric effect, we had to think of light in a new way. Why not take a heuristic approach and develop a model designed to solve this problem? In 1900, Max Planck had taken a similar approach to solve the seemingly intractable problem of blackbody radiation. The solution required that he make the assumption that radiation was emitted from a glowing solid in tiny bursts, which he called quanta. The quantum hypothesis, while not at all appealing to Planck, allowed the experimental data to be explained.

Einstein assumed that perhaps light, too, could be viewed as composed of discrete packages of energy, or photons. And the energy of these photons depended on the frequency of the light. Relatively low-frequency radiation, such as red light, had low-energy photons, whereas high-frequency radiation, like ultraviolet or x-rays, had photons with much higher energies. Using this model, it was possible to imagine photons of red light striking a zinc cathode and just not having sufficient energy to dislodge the surface electrons. Ultraviolet light, on the other hand, was composed of far more energetic photons, and these did have what it took to dislodge the zinc electrons and send a current streaming out of the zinc cathode.

The model worked. It explained the results of past observations, and it passed new tests devised specifically to challenge it. But it also revised the way that we think about an important feature of our world. Light was no longer just a wave as it had been portrayed since 1805, when the physician and Egyptologist Thomas Young conducted his slit experiments on interference. With Einstein's work, it became necessary to think of light as exhibiting a particulate nature also. Not long after 1905, it became clear that light had a dual nature: it was a wave and it was particle. How you thought of it depended on the context.

With the work of Max Planck on blackbody radiation in 1900, and then Einstein's 1905 paper on the photoelectric effect, a new science was coming into being. It was quantum mechanics, and, along with relativity, it would forever change our image of the physical world.

Einstein had given us more than a deeper theoretical understanding. Much of practical significance came from the photoelectric effect: garage-door openers, smoke detectors, night-vision goggles. But at the Hotel Glocke, I was thinking, for some reason, of photovoltaic cells, of sunlit fields and deserts, of the daylight breaking over Nellis Air Base in Nevada, over the gentle hills of Bavaria and Portugal, of vast arrays of silicon cells laid out awaiting the sun, tracking it through time, through the long day, and of the gathering of photons falling onto the dark plates with their doped-in phosphorus and antimony and arsenic, and the weakly bound electrons about to be dislodged, just as Einstein imagined here in Bern, to be sent as current, as pure electricity from the sun's rays

without all the commotion, the steam and noise of coal and oil—so
nineteenth-century, so classical and disruptive of this thin shell of air—
with just the silent turning and tracking as the Earth turns, as it warms
itself before the great fire. This vision of photons and electrons—light
and matter interacting in ways that we never understood until then,
until Einstein thought one day how, just maybe, we could think of light
as something other than a wave—would in time light the cities with a
softer glow and send the engines spinning wheel to wheel, and all without the carbon that darkens our dreams and sends the oceans closer to
our doors and the glaciers retreating far into distant hills.

❧

The special theory of relativity changed forever the long-accepted
view that time, length, and mass, the very fundamentals of the physical
world, were invariant from one observer to another. As Newton had
written, "Absolute, true and mathematical time, of itself and from
its own nature, flows equally without relation to anything external."
Einstein had shown, through his famous thought experiments, that time
and duration were, in fact, relative and dependent on the observer.
There is no absolute Newtonian time equitably flowing like an untroubled stream through the world. Nor is there absolute mass. An electron
moving at nine-tenths the speed of light in an accelerator is more massive than an electron at rest or an electron attached to a droplet of oil
slowly sinking through air. And the jet liner in flight, as measured from
the ground, is actually shorter than the same jet liner on the runway. In
1905, these were difficult ideas to comprehend, but by 1912 they had
largely been accepted by the physics community.

Also, there was the separate paper on the relationship between
mass and energy, the one that contains the most famous equation in all
of physics. From 1905, in the city of Bern, there extends a long causal
thread that winds its way through Munich and the work of Meitner
and Hahn and Strassman, and then on through the drylands and mountains of New Mexico and into the heart of the Japanese empire and the

ships of MacArthur anchored in Tokyo Bay. Not long ago, in California, I saw an exhibit of Richard Avedon's photos, and among all the politicians and writers and artists and poets there was only one scientist—the man from Los Alamos, J. Robert Oppenheimer, who knew how the equation had changed everything. In Santa Monica, across from the RAND Corporation, there is a public sculpture made of links of chain. It rises thirty feet in the air and is shaped like a mushroom cloud. It is called "chain reaction" and the inscription reads: "This is a statement of peace. May it never become an epitaph."

I had a good breakfast at a restaurant near the Glocke and then I left the city having seen what I had come to see. From the train window, the river Aare appeared to run swift and cold with all the fresh snowmelt from high in the Alps.

❧

I stayed only a year in Washington. During that time I read not a single word of Schilpp's book on Einstein. I made the varsity baseball team, played in Maryland and Virginia and around the South. My father would drive his old Pontiac down from Pittsburgh to see the games. I stayed in touch with McMullen, who wrote to me about college and his new Mustang and his nights at the drive-ins and the young ladies he was meeting. I wrote long letters to Mary and to my family. I still have drafts of some of those letters. But I cannot read them today without wincing at their certitude, their youthful preachiness. In the spring, my lab instructor, Sally Sheehan, told me I would make a good chemist and that I should give some thought to changing my major. I thanked her for the advice and told her I had really enjoyed her course.

Schilpp's book stood untouched on my shelves for many years. When Barbara and I taught our course on scientific revolutions, I took it down and reread what Einstein had written about himself and what Bohr and others had written about him. The book had been published in 1951. The receipt from Walleck's bookstore in Pittsburgh for $1.37 was where I had left it long ago, somewhere near page ten.

Pittsburgh
and Paris

You never forget the sound steel makes against shoveled coal, or the resistance of the shovel to the metal as you dig beneath the large chunks. Nor do you forget the sound of coal against the chute. And then there is a rush of cold air into the house that you welcome as you stand in front of the open furnace with its heat and redness and the roar of the flame. In our house I learned that this is where warmth comes from—from the combustion of coal—and it was my responsibility to see the furnace supplied and stoked and to remove the powdery white ash as it settled in steel buckets.

In Pittsburgh, fire was all around us. You had to drive only a few miles to the nearest mill, which lay just south of the city on the river. At night the mills made their own sky, filled with smoke clouds and redness, and the long slag heaps to the west cast a flickering glow on the horizon. At times, you felt like you might have been standing on the plains of Mars.

The eerie mills were the city's living heart. They made the air you breathed, or at least the part that you didn't want to breathe, turning it dark with smoke, brown with the taint of oxide gas. They made the rivers run iridescent with oils along the shore and the smell of creosote and rotting fish. Because of them, the bars of the city never slept, serving up shots and beers with the shifts that ran day and night. There were a thousand small factories that fed off the mills, and many of them, like FESCO, where I worked summers, had dispensers of salt tablets along the assembly lines so you could keep your sodium in balance while you sweated and hung sinks on the line before they were enameled. Dewer Spring, Dravo, Press Car Steel, even the bakeries, like Mancini's, where you ordered stick loaves by the dozen, were there because of the mills and the tempo they set for the place.

My mother would recall how in her day the sky rained with soot, and there would be times when downtown was dark at noon and the gas lamps cast haloes of gloom into the midday air. One of my favorite photographers, W. Eugene Smith, who later became famous for his work on mercury pollution at Minimata Bay, Japan, was so intrigued by the industrial landscapes of Pittsburgh that he extended a three-week magazine assignment there into many obsessive months. He shot seventeen thousand pictures of Bessemer converters, human faces, and the grime and life of a city built on burning. It is little wonder that Charles Dickens called it "Hell with the lid off." Henry Miller considered it the ninth circle of Hell on Earth.

Miller had just returned from Paris and was about to begin a road trip across America as a way of rediscovering his native land. He had standards of comparison. I had none. Being born into the place, it was all that I knew, and many of my early images were of things burning or about to be burned, heating or about to be heated, glowing or about to glow. The furnace alone, which I tended from a young age, seemed as mysterious and challenging as the stars. It was easy to see that something was being changed, that the coal that had rolled so noisily down the silver chute into the basement was being reduced to a puffy ash. What happened to it? Why did it burn?

We have always burned things—at the mouths of caves, on the banks of streams and the shores of lakes, in the remoteness of the polar night, in our furnaces and factories and rocket engines set for the Moon. As much as anything else, we are the ones who create conflagrations, whole forests set crackling and luminous, drenched in

A replica of Lavoisier's laboratory.

fire. And yet we have never really understood burning until recently.

Despite the reputation it has acquired over the years, chemistry is a science not just about laws, principles, and symbols on a chart, but about the imaginary, hidden, but nonetheless real world. The substratum that lies below our sight and touch, our hearing. And to enter that world, to walk among its creations, its molecular architecture of pyramid and tetrahedron, is no less an adventure than to follow Calvino into Moriana, one of his "invisible cities." In a way, this adventure in chemistry begins with Lavoisier, in the very real city of Paris.

If you think about the process of burning for a minute, you realize it is always the same. The log in the fireplace, the wax candle in your room, the charcoal in the grill, the old fires in the mills of Pittsburgh that I remember from youth, even the gasoline fire I once saw in an exhibit not long ago, racing along the floor toward the fire artist who had lit it. In each case, burning seems to involve a release, a giving-off of something. The flame, the dark smoke, things rising, losing themselves in air.

In the middle of the seventeenth century, the idea that burning somehow happened exactly the way you see it, that something is set free, was proposed by Johann Becker. Becker taught that all flammable objects contain a fatty substance called *terra pinguis* or fatty earth. It was this "fatty earth," dispersed throughout the body of that coal I once shoveled, that was sent into the air until it was gradually exhausted in the process. A little later, George Stahl, a Luthern priest and professor of medicine at the University of Halle, renamed *terra pinguis* "phlogiston," which is the Greek word for flame. Stahl developed the phlogiston theory into one of the great organizing principles of early modern chemistry.

But this theory went well beyond the large class of familiar objects that we see burning every day. It also explained a process familiar to alchemists or to anyone who has ever witnessed the rusting of iron, who had seen swords and wagon wheels or their favorite Chevy grille crust over with brown and red scabs, become friable, and turn to dust. This unhappy process, which in our own day consumes countless million of dollars, was called *calcination*. Here too, according to the theory of Becker and Stahl, phlogiston was involved.

※

I could not have asked for more pleasant days. It was May 1995, in the city of Paris. The air in the gardens was sweet. The stalls along the Seine were filled with tourist art—the Eiffel Tower, Notre Dame, Montmarte with its white-domed Basillica—and paperbacks by Rilke and Camus and Hemingway and Milan Kundera, a whole avenue of kiosks and books along the Left Bank. I wanted nothing more than to spend an afternoon in the cafés.

On a napkin, I sketched out a reaction for how the old phlogiston theory may have worked. I thought of a perfect silvery nail in the way that Becker and Stahl did. It is made up of only two things, iron and phlogiston. When the nail rusts, phlogiston is given off and the steel turns into what they called a calx. On the napkin before me, the reaction looked like this:

Perfect silvery nail = calx (rust) + phlogiston

If you want to reverse the reaction and get the perfect silvery nail back, all you need to do is heat the calx with something that has lots of phlogiston, like charcoal. And the reaction on my Parisian napkin looked like this:

Calx + phlogiston = perfect silvery nail

Again, in both of these reactions, the perfect silvery nail is considered a kind of compound made up of iron and phlogiston. It is a rather neat idea.

I recall that there was a professor of chemistry at Berkeley who began his course by teaching the phlogiston theory. He would mount arguments in its defense and leave the hall full of hundreds of twentieth-century students wondering, at the end of the lecture, what age they were living in.

As with Aristotle's views of the Earth, there were serious problems with the phlogiston theory. As Thomas Kuhn would say, there were "anomalies." They tend to slowly erode a theory from below until it eventually collapses. As it turns out, no matter how lofty or inclusive or faultlessly pedigreed a theory is, in the end it is vulnerable to observations and experiments. When he was working on his *Principles of Psychology*, William James wrote to his brother Henry, saying, "I have to forge every sentence in the teeth of irreducible and stubborn facts."

Amid stale smoke, lounging dogs, and a few silent patrons, I ordered another beer and thought about one of these problems: weight. If you burn a log in your fireplace, or a lump of coal in the furnace as I did in Pittsburgh, it is clear that the ash weighs much less than the wood or coal from which it was derived. This makes sense since phlogiston is lost in the process. But in calcinations, where phlogiston is also lost, the resulting calx actually weighs more than the metal from which it was formed.

In the first case, phlogiston appears to have a positive mass; in the second case, it has negative mass. How can it be both? Antoine Lavoisier, who in this city in the late eighteenth century had the finest balances anywhere in the world, realized that this was a major inconsistency in the phlogiston theory. There were other problems, too, and in 1885 he called phlogiston "a vague principle, which is not strictly defined and which consequently fits all the explanations demanded of it." It is, he concluded, "a veritable Proteus that changes its form every instant."

Across Europe at this time, there was a parsing of air, a teasing apart of one gas from another. It was becoming clear that air was not an elemental substance, as Aristotle had thought, but a complex mixture of gases having unique properties. One of the champions of this effort was the preacher Joseph Priestley. In 1774, he had directed sunlight onto mercuric oxide and released pure oxygen gas, which he soon realized had properties that were different from air: candles burned more vigorously in this isolated gas, and there was a clear physiological effect to breathing it which Priestley—the proprietor of the first oxygen bar?—clearly enjoyed. A supporter of the phlogiston theory, he never understood the chemical significance of the gas he had liberated.

Priestley was a Unitarian, a critic of the English political system, and a supporter of the American and French Revolutions. He was eventually forced to flee his home in Birmingham, when a mob set fire to the family house and destroyed his laboratory. An outcast, he arrived in 1791, at the age of sixty, in Northumberland, Pennsylvania, on the banks of the placid Schulkill River in the midst of trees and nothing much else. Today you can visit his laboratory, view his study, and learn that he was the inventor of carbonation, a process that enhances the quality of the very beer I was sipping at the café.

❦

Every museum of science that I visited in Europe had a display devoted to Lavoisier's laboratory. The equipment looked anything but antique and, in fact, some of it was beautiful—works of art in glass and copper that inspired respect and admiration. There were the retorts and the finely crafted balances, a great lens that had been constructed for concentrating sunlight at the French Academy of Sciences, and the ice calorimeter he had used for measuring the heat released in a chemical reaction. It was clear, standing before these objects, that finally chemistry was moving beyond its alchemical past into a future of precise measurement and quantitation. It was through this equipment that the law of conservation of matter came into being. In Lavoisier's words of 1785, "Nothing is created either in the operations of art or those of nature; and it may be considered as a general principle that in every operation there exists an equal quantity of matter before and after the operation."

The exhibits everywhere celebrated Lavoisier as a brilliant experimentalist. But before the experiments came the hypotheses, the crude and then more refined guesses, the thoughts that began mysteriously— who knows exactly how—in the mind, that challenged convention.

The lumps of coal fed into the fire and set ablaze had nothing to do with phlogiston, that troublesome, nonexistent chimera. Its burning involved, as Lavoisier would show, the uptake of oxygen from the atmosphere—the gas that Joseph Priestley had isolated but had not really understood. And it involved the release of carbon dioxide and water. And rusting, too—the perfect silvery nail inscribed on my napkin that delightful afternoon in Paris—was for Lavoisier a process of uptake, of oxygen from the atmosphere combining with metal to form the rust, the iron oxide as we now know it.

In one of his scientific papers, published in Paris in 1777, you get a glimpse of Lavoisier's mind and his careful experimental method. He reports how he added four ounces of mercury to a flask filled with air, sealed the flask, and heated it to just below the boiling point of the metal. He observed the flask daily, reported on the red flecks he saw forming on the silvery surface of the mercury. After twelve days, he tells us, he removed the flask from the fire, let it cool, and then studied the properties of the remaining air. He observed that it had been depleted in oxygen by "one sixth its volume," and that it "extinguished flames" and "killed rapidly animals that were placed in it." It was, he said, in a "mephitic state." He realized that all of the oxygen had been removed and had formed a calx with the mercury, what today we would call mercuric oxide. In effect, the broad group of calcination reactions was finally explained correctly in terms of oxidation.

Lavoisier's shift away from phlogiston—what Kuhn called a paradigm shift—was as important for the development of chemistry as Newton's or Galileo's ideas were for physics. It set the stage for Dalton's atomism and for the long procession of ideas that passed through Mendeleev and Rutherford and Bohr and through all those who imagined and refined the invisible world that lies below the tables and chairs of our experience.

❦

I never walked so much as I did that week in Paris. I visited the church of Saint Germain-des-Prés, where a handout in English read "The oldest church in Paris, consecrated on 21st April 1163." The remains of the seventeenth-century philosopher and mathematician René Descartes lie there. In the small chapel, to the right of the apse, where it was dark and forgotten, I lit two candles in his memory. I visited the Pantheon and the tomb of Sadi Carnot, whose idealized heat engine led to one version of the second law of thermodynamics—a law that Ludwig Boltzmann would later interpret in statistical terms. Harv once told me in West Lafayette that he never liked the Carnot engine. "It's a damn steam engine," he said, with complete contempt. But I found it ingenious. It seemed, like so many of those Platonic idealizations in science, a perfect way to distill the essence from something that was impossibly complex.

At the Pantheon, I looked into the vast space where once Foucault pendulum had swung with such "isochronal majesty," and had demonstrated for all to see the turning of the Earth. I imagined Copernicus, Galileo, and Kepler standing nearby, vindicated finally, smiling and nodding in approval. And at the Jardin des Plantes, where the young Lavoisier had first witnessed the magic of chemical demonstrations, I rested amid an Alpine garden and tried to catch up on my notes. Later I attended a lecture and slide show at the Institut Curie. I wrote in my journal that "there was a black-and-white photo of the old, white-haired Marie, standing on a hill with Albert Einstein. Nothing there but the two of them and the sky. They looked as though they were the only two people on earth."

Before I left the city, I knew I would have to visit a place I had resisted for days. It was the Place de la Concorde. This is perhaps the largest public square in Paris. It stands at the eastern end of the Champs-Élysées. During the Revolution it was the site of the infamous guillotine.

If you look up the broad avenue, it is possible to think how it must have been two hundred years ago. To the south, there were then great trees, which stood in perfect rows now crowded with buildings. Just beyond, where the Arc de Triomphe now rises, there were farm fields and little farmhouses. The Seine to the east ran blue. North of the Louvre, much of the city, with its red rooftops, stretched into green plains.

On Thursday, May 8, 1774, Lavoisier and the twenty-seven other condemned members of the *Farmers General* (a group of tax collectors) were led—hands bound behind their backs—to the foot of the guillotine. He was fourth, behind his father-in-law, M. Paulze, and when he ascended the wooden stairs to the platform, he would have looked out on evening falling over the pastoral landscape. They say it took only thirty-five minutes for the executions to be carried out. Afterward, the bodies were stacked like so much firewood on wagons and the heads were put in wicker baskets. A procession escorted the dead to a common gravesite outside the city, called Errancis, "the place of maimed persons." In 1859, to make room for road construction, the remains were transferred to the catacombs beneath Paris. The mathematician Lagrange would say after Lavoisier's execution that "it took but a moment to cut off that head, though a hundred years perhaps will be required to produce another like it."

When he was executed, Lavoisier was a wealthy man. He had earned more than a million dollars a year, but the earnings came from his activities as a tax collector—a reviled profession. Wealth had given him the resources to build one of the great laboratories in Europe. But it also made him an obvious target in Revolutionary France. I had wanted to see how people today felt about his execution, so I asked around at the hotel and in the cafés. Was there a sense of national guilt? Of intellectual waste? After all, it was as if England had hanged Isaac Newton in London, or America had marched Benjamin Franklin before a Philadelphia firing squad. But of the people I spoke with, no one remembered anything about him. Paris did name a street after him, and they erected a statue in his honor. Unfortunately, they placed on it the head of the philosopher Condorcet. No one noticed for a hundred years.

Time and art, however, have worked in Lavoisier's favor. At the Louvre, I went in search of the portrait that Jacques-Louis David had made of him. I had seen it first at the Metropolitan Museum of Art in New York. Because it was formal neoclassical portraiture I had been prepared to dislike it; instead, it had amazed me. Its provenance had been complex, from Lavoisier's wife, Marie-Anne, to her great niece, to John Rockefeller, to the Rockefeller Institute, and finally, in 1977, to the Metropolitan.

Lavoisier's absence from the walls of Parisian art has been more than compensated for by his presence in nearly every Chemistry 101 text ever published. My old college text features a small black-and-white reprint of David's famous portrait of him and Marie-Anne in their study.

David's painting has become an icon of the chemical profession. There is nothing quite like it in the realm of scientific portraiture. It shows a young woman in a formal white gown, braided hair falling down her back, slightly bent over a writing table. She looks outward from the canvas toward the viewer. Seated at the table, quill pen in hand, is a man dressed in black, his leg extended into the foreground, touching the red velvet of the table, his face and his gaze turned upward toward the woman. Her left hand rests lightly on his shoulder, and there appears to be an easy bond between the two. The instruments of the chemical profession can be seen on the desk and nearby on the floor: a barometer, a bell jar, a device for accurately measuring the volume of a gas. I was captivated by it, by the vivid colors—the blue silk trim of the dress, the crimson velvet—and by the sheer size of the canvas, which is nearly nine

*Jacques Louise David's portrait of Antoine Lavoisier
and his wife Marie-Anne.*

feet by six. It is signed at the bottom by the artist and dated 1788. A mere six years later, Lavoisier would ascend the scaffold to his death.

※

Like so many travelers, I had fallen in love with Paris and had not wanted to leave. But unlike W. Eugene Smith, who had taken such a liking to the grimy, romantic city of my birth, I did not have the luxury to stay. And so I took in my last views of the city from a *bateau mouche* on the fresh breezy Seine. I thought of all I had seen and all I would never see in this inexhaustible place of science and tragedy, and how nothing could eclipse what Lavoisier had taught us about the very fires of our beginning.

121

Juneau

One terrible night in 2006, when I was in the ICU, in pain and drugged on morphine, silent nurses coming and going to take blood in the dim light, Camus's words came to me. I was more or less reliving my life the way you do when you're not sure how much of it you might have left. Perhaps Edgell's lecture on Boltzmann provided a clue. Perhaps the plains and skies of Indiana. Perhaps Harv and Buckley and Rita, or the train through the night after Kennedy's death. Or the Buhl, with its pendulum and giant sextant, its smooth ellipses on a chalkboard. Or something else, something far more subtle, the vague feeling the stars gave when, as a child, I stared at them from the attic window. In autumn, the crunch of leaves underfoot, the winter snowfalls, the moods of those spring nights. At times there were little flashes, like something you see at sunset over an ocean, little insights that came and went and were forgotten. The kind woman from the church, who had just brought me Communion, said, "Are you planning to die?" I was lucid enough to say, "Yes, but not tonight."

Some months later, I had been through all the tests. The bone marrow samples and blood had been taken and sent to the Mayo and Cleveland clinics. A diagnosis had been rendered—stage four non-Hodgkins lymphoma. There followed two surgeries, eight rounds of chemo, the usual baldness, a loss of fifty pounds. Whole months went by that year when I didn't shower for fear that I would catch a glimpse of my legs, which had become thin and formless as the winter branches beyond the window. Most of the time I lay either in bed or on the sofa. Even when friends came, I barely moved. Wanda kept a journal, recorded my temperatures, drove me to the hospital for appointments. I read as little as I could about the illness, but I knew the chemicals they were giving me were lethal. And I knew their structures, down to the last carbon atom.

When it was over and I had regained some strength and the positrons in the PET scan were seeing nothing but the usual nodes, my friend Newell, who had come to visit almost twice a week, asked Wanda and me to go on a cruise. "Susan and I have always wanted to see Alaska," he said, with that infectious enthusiasm he brought to his plans. Alaska had never been a dream of mine. Cruising the seas in a floating hotel was likely to be the kind of nightmare experience David Foster Wallace had described in his 1997 collection of essays, *A Supposedly Fun Thing I'll Never Do Again*. But Newell had been there for me through an awful year. So I agreed to go.

❦

Before Priestley and Lavoisier had done their work on gases and had suffered at the hands of angry mobs, the Scotsman Joseph Black was calmly pursuing his chemical studies at Glasgow and Edinburgh. Black was heating solids like magnesium carbonate and collecting the gases that evolved. Collecting gases had been made possible only recently by Stephen Hales, who invented something called the pneumatic trough. It was simple and ingenious, like most revolutionary inventions, and it made the eighteenth century, in chemistry, the Century of Gases.

I recall using Hales's device, in updated form, in my high school chemistry course. There was a florence flask (named after Galileo's city), round and sparkling with a slender neck, and black particles of manganese oxide at the bottom. McMullen and I worked together on this. From the flask we ran a thin tube over to a pan partially filled with water. The tube, snuggly fitted in a cork in the mouth of the florence flask, was the conduit that ushered the gas over to the water, where we could collect it in test tubes. It was a marvel to just watch it slowly dis-

The Mendenhall Glacier as seen from the visitor's center.
A sign calls attention to the glacier's rapid retreat.

place the water in the test tubes as it bubbled languidly with its hypnotic *blurp blurp* sounds. Neither of us had ever heard of Stephen Hales.

In this way, we collected pure oxygen, just like the chemists of the eighteenth century did. We did all the usual tests: dipped lighted splints into test tubes and watched them flare and turn the wood to ash; put spirals of heated magnesium into the pure gas and let it blind us for an instant with intense light. "Wow," McMullen said, "I've never seen anything like that." Neither had I. Neither had any of us. "If this is chemistry," I thought, "it's more interesting than I had imagined." I carried these images of burning with me for a long time, through the autumn when the buckeyes were dropping shiny and smooth as gemstones from the trees, and through the gray winter, when the snows came and transformed the land. On those days, the hills of Pittsburgh seemed steeper and the rivers broader and clearer than usual.

Joseph Black used this technique to isolate that subtle but important atmospheric gas, carbon dioxide. According to Aaron Ihde, in *The Development of Modern Chemistry*, Black was interested in using mag-

125

nesia alba (magnesium carbonate) as a way of dissolving kidney stones. In connection with this project, which ultimately failed, he looked at the chemistry of magnesium carbonate in some detail. In a series of experiments he noticed that by heating the magnesium salt he could cause it to lose weight. He concluded that an "air" of some kind was being lost, so he collected it. What he found, however, was not ordinary air. Its properties were vastly different. It extinguished flames. It killed the small animals he placed in it. And it readily turned a solution of quicklime (calcium oxide) turbid and cloudy. He called this new substance "fixed air." Today we know it as the fourth-most-abundant component of the Earth's atmosphere, carbon dioxide.

Black learned that he could generate carbon dioxide by heating limestone (calcium carbonate) and by treating carbonate salts with acid. He also showed that exhaled air was rich in this gas. And he was the first person to realize that carbon dioxide was present as a small component of air. He understood that common air was a mixture of gases and not the elemental substance that Aristotle had claimed it to be. There was no such thing as elemental air. Black's work, done in the 1750s, paved the way for the research of Joseph Priestley and Antoine Lavoisier.

※

The cruise ship *Volendam* inched its way up the inside passage from Vancouver north. Along the shore there were vast stands of pines. From the upper decks you could see the sun going down, and there were shoots of red and orange on the horizon and small islands in the distance. In the evening there were floor shows, and the casino was bright and crowded. At eight, we went down to hear the string quartet play Bach and Mozart. There were only a few people in the room. I thought of Boltzmann and how he must have heard these same notes in Vienna more than a hundred years before.

When the ship docked at Icy Strait Point, some of us boarded the open-air tenders and were taken ashore. There was a hundred-year-old salmon cannery near the dock that, from the outside, looked like something out of Steinbeck. Inside there was a museum and a gift shop and various art galleries, where I purchased a somber photograph of the Northern Lights. The lights in the picture glowed blue in the distance behind a small Eastern Orthodox church. The photograph looked like it had been taken somewhere near Sitka, which was once the capital of Russian Alaska. A local paper that we picked up in one of the shops said the village of Hoonah was a mile and a half from the cannery. Hoonah had a population of 880, most of whom were from the Tlingit Tribe.

We hiked up the road to the village. It was chilly even though it was June so we wore our jackets and knit hats. We passed an old graveyard with weathered crosses that seemed to be toppling among the tall ferns. A blue and red totem pole rose against the gray sky; beyond you could see the ship anchored in still water. There was a wooden building with a big sign that read HOONAH PUBLIC SCHOOL HOME OF THE BRAVES. On the rooftops, there were too many bald eagles to count.

We stopped at the Landing Zone Restaurant for lunch. Susan wanted to know how the town had gotten its name, and learned from one elderly native that when he was a kid the Strait was filled with ice far into the summer. Gradually things had changed, and now it was clear. You could see that by looking out the window, which was right across the room. There was no ice at all.

The next morning we docked at the capital, Juneau. The place looked like a small village nestled between the sea and the steep mountains, which rose to heights of four thousand feet. From the deck where we drank coffee, you could see the Mount Roberts tramway climb into gray cloud. The city was named after a gold prospector, Joe Juneau, but I preferred the old Tlingit name, Dzantiki Heeni, which means "river where flounder gather." The city had a population of about thirty thousand and was in the midst of the northernmost rain forest on earth.

When we finally got off the ship and went into town, it was sunny. People were smiling and nodding to one another. It had been raining for weeks.

Wanda wanted to see one of the gardens that had been advertised in a travel brochure, so we took a bus into the hills and walked toward the park. We were in steep terrain, and to get around they drove you on canvas-covered golf carts fitted with powerful motors and, we hoped, brakes. The gardens were unlike anything I had ever seen. Trees had been uprooted, turned upside down so their roots were high in the air. In the roots there were baskets of trailing flowers that were pink and violet against the blue sky. The paths that we took through the Tongass Forest were narrow and steep and passed near ponds and waterfalls and precipitous cliffs, until they reached the summit. There on a platform you could view the great sweep of the Mendenhall Valley and the broad river that flowed from the melting ice fields. Beyond the river there were snow-capped mountains. Back at the visitor's center was a book where people wrote their thoughts: "So beautiful it made me cry," said one entry; or "Makes me want to go home and dig in the dirt." I think Newell enjoyed this, but not as much as he expected. Over the years he had become a connoisseur of fine gardens around the world and was not so easy to please.

On the platform overlooking the valley, I asked my usual question. Even though the drivers were young they said they had noticed some changes in recent years. One complained that the skiing wasn't as good as he remembered it as a kid. And another said it seemed that the river down there was stronger and swifter. Susan had gotten similar stories from the people she spoke with.

We caught a bus for the Mendenhall Glacier, which was really what I had come to Juneau to see. We were dropped off in a parking lot and walked for a few miles along a paved road through the Tongass National Forest. Everywhere there were ferns and thick stands of trees. At first sight of the glacier, I took pictures of Bill and Susan in their heavy shirts against a backdrop of the distant ice fields. When we arrived, you could see a large lake and tourists standing at its edge, and then the glacier sloping downward with its face in the water. I was surprised by how much rock debris had accumulated on its surface. It was not like the silvery white sheets I had become familiar with in Antarctica.

Inside the center, there were the usual touristy things: a store that sold maps and books and stamps and various trinkets; a lecture hall and theater; a small cafeteria. But my attention focused on a single plaque. It was positioned before a large semicircular window that looked out onto the blue glacier and its crystalline lake.

The plaque read in bold print: IN 1935 YOU COULD TOUCH THE FACE OF THE GLACIER FROM HERE. I envisioned myself reaching out and feeling its raw coldness against my hand. It must have been thicker then, towering over the visitor's center, a huge force present to me now only in my imagination. Now it was more than a mile away. Its snout seemed small and insignificant compared to what it had once been. Large chunks of ice that had calved away from it were floating in the lake.

I learned that the Mendenhall has been receding since the mid-eighteenth century. The balance between snow accumulation in the Juneau Snowfield and melting at the glacier terminus is now biased toward the latter process, so the glacier retreats. It is a seemingly slow process, but standing there at the window I could see it in dramatic form: 1935 to the present. A time span we can easily comprehend. Our grandparents, our parents were alive. Perhaps we were too. And yet something momentous has occurred, a feature of the Earth's permanence has shifted, has left us wondering about time and stability, about things we took for granted.

This bias toward melting is not unusual. Everywhere on Earth the glaciers are retreating. You can see it in the photographs of the Qori Kalis Galcier in Peru, taken fifteen years apart, or in the Upsala Glacier in Patagonia between 1928 and 2004, or in the glaciers of the Swiss Alps and the Tibet Plateau. Lonnie Thompson, who has studied high altitude glaciers around the world, has photos of the Southern Icefields of Kilimanjaro from 2000 and 2006. Kilimanjaro is the highest mountain in Africa. Thompson's photographs show that in a mere six years there has been a dramatic wasting of ice from the summits. In one photo, Thompson leans against a tall splinter. In that area of the great mountain, it is the only ice that remains.

When we got back to the ship, we had dinner and talked about glaciers. Most people had never seen one. They were surprised by the beauty of the Mendenhall, by the fact that it seemed alive, with its sounds and the waters that streamed from its face. There was great anticipation about the Hubbard, which was the next stop.

Wanda and I went up to the Crows Nest. We drank Wild Turkey and looked out at the sea and the fading light on the horizon. It was nearly midnight. Things seemed to me to connect, mysteriously, through time as though a great tapestry you can't understand were taking form. There, was a soft chill in the air and I raised my hand to order another drink.

With the method of Hales and the work of Joseph Black, we were beginning to understand the complexity of air. Except for argon, which was not discovered until 1894, the four major atmospheric gases had been isolated and studied by the end of the eighteenth century: nitrogen, oxygen, argon, carbon dioxide, in that order of abundance. I often ask my students what the atmosphere contains, and most will say it is predominantly oxygen. But were this true, every lightening strike would bring down the house and turn the forests and fields to tinder. Some students will say that carbon dioxide is plentiful, when it is really only a trace, a fraction of a percent. Primo Levi says of this gas, "which constitutes the raw material of life" and "is the ultimate destiny of all flesh," that it is a "ridiculous remnant, an impurity."

Alas, hardly anyone thinks of argon, that perfectly inert element whose atoms are round as basketballs. Its very name means "inactive." In my own work, I came to respect it for its utter indifference to the molten nitrate salts I ran it through in the lab. At a surprising 1 percent, it is three times as plentiful as carbon dioxide. But, molecule for molecule, nitrogen surpasses all of these, and makes our atmosphere reasonably pleasant and not as explosive as it might otherwise be.

If the eighteenth century marked the Age of Discovery and the birth of real chemistry, it was the nineteenth that provided recognition of what all this meant to us. Early on, you could calculate, using straightforward physics, that a planet the size of the Earth, with the Earth's distance from the Sun, would be a cold and inhospitable place if it were simply a smooth rock—an airless stone circling in space. So why is it warm? As early as the 1820s there was speculation that the atmosphere, this "fragile seam of dark blue light," might be responsible. In the 1820s, Joseph Fourier speculated that while visible light from the Sun could easily penetrate the atmosphere, the outgoing radiation—quite different in character—might be trapped. John Tyndall's laboratory experiments in the 1850s showed that both water vapor and carbon dioxide were capable of absorbing this radiation and thus causing a gentle warming of the planet.

Toward the end of the century, the Swedish chemist Svante Arrhenius published an important though largely forgotten treatise on the role that carbon dioxide—Joseph Black's "fixed air"—has exerted on the Earth's climate over time.

Before he ever wrote his controversial papers on climate, Arrhenius had done several memorable things. No one could understand why ordinary salt would not conduct an electric current. Nor would pure distilled water. Yet if you mixed the two, if you added salt to water, the solution that resulted was an excellent conductor. To explain this, Arrhenius had said that the salt, once it touched water, dissociated and formed ions. A little table salt, sodium chloride, disappears from view when dropped into a glass of water, and in so doing forms bits of charge, the positive sodium ion and the negative chloride ion, which swim in the glass, just as they move in the sea off Alaska. Arrhenius's chemistry had its invisible entities, far below the power of our senses to discern. For this work, communicated in his paper "On the Dissociation of Substances in Aqueous Solution" in the typically unemotive language of scientific journals, Arrhenius received the 1903 Nobel Prize.

In addition to this, he developed a new definition of acid, which he considered to be any substance which generates hydrogen ions in water. This definition includes the gas carbon dioxide, which is composed of only carbon and oxygen and contains no hydrogen at all. But when added to water, carbon dioxide will form carbonic acid, which in turn dissociates to give a hydrogen ion. This is why a drop of pure, uncontaminated rain, interacting only with the air around it, turns out to be slightly acidic.

What interested me, after seeing the Mendenhall Glacier and its rapid retreat, was what Arrhenius had said about carbon dioxide in his 1908 book *Worlds in the Making*. Directed to a general audience, it was prescient, though his major concern was not precisely our own. He was focused on the causes of the Ice Ages, which he approached through the lens of atmospheric chemistry, incorporating the works of Fourier, Tyndall, and others. Like Tyndall before him, he emphasized the role played by water vapor and carbon dioxide in the absorption of infrared radiation emanating from the Earth's surface. Importantly, he had access to the works of Frank Washington Very and Samuel Pierpont Langley, who had made infrared measurements of moonlight at the Allegheny Observatory in Pittsburgh.

This historic facility, which stands on the north side of the city high above the Allegheny River, is still an active research center. It is the only observatory known to me where a telescopic lens was stolen and offered up for ransom. The ransom was never paid by the director, who happened to be Langley himself, but the guilty "lens-napper" became sufficiently paranoid in its company to dump it in a trash can in nearby Beaver Falls. The scratched lens was later reground and, as a result of this optical reworking, became far better than it had ever been. When I was growing up, everyone knew the observatory mainly as a great place to "park."

❧

The interaction between carbon dioxide and certain frequencies of light is a favorite topic in physical chemistry courses. I cannot recall whether Walter Edgell, back in Indiana, ever mentioned in his ode to Ludwig Boltzmann that Svante Arrhenius had actually worked with the master as a student in Graz. But Edgell, later in the course, did discuss the spectrum of carbon dioxide. As it turns out, the light coming back from the Earth has just the right energies, the right frequencies, to make the carbon dioxide molecule move in special ways in a kind of molecular yoga. To one frequency, it stretches, elongates itself along a straight line. To another frequency, it bends, with the central carbon atom poking itself up above the line of the two oxygens. These are the warm-up exercises for what will come later.

Arrhenius reasoned, a hundred years ago, that the more carbon dioxide molecules there are in the atmosphere, the more absorption there will be. But more absorption leads to a warmer atmosphere and a warmer earth. The argument is straightforward. Using a law developed by Stephan and Boltzmann, he calculated how much warming would occur should the carbon dioxide in the atmosphere double. His estimate was 5 or 6 degrees Celsius. This is remarkably close to our best present-day estimate of 2 to 4.5 degrees. And we are nearing this doubling—not in the distant three thousand years that Arrhenius predicted, but in decades.

Svante Arrhenius, who was Swedish, believed that warming might just be a good thing—like the people to our north in Akron or Cleveland today. So did his colleague, the physical chemist Walther Nernst, who thought it wouldn't be bad to set a few useless coal seams ablaze just to help things along.

※

By the end of the twentieth century, global average temperature had increased, just as Arrhenius said it would, by about 1 degree Celsius. Most of the world's mountain glaciers were in retreat. The snows of Kilimanjaro were nearly gone.

For me, Pittsburgh had always been the "City of Burning." But there are now cities of burning everywhere—Los Angeles, Mumbai, Beijing—more than Arrenhius could ever have imagined, spread across the globe.

There can be little doubt about the causal threads that weave their way from the mines and oil refineries through the factories, cars, planes, and cruise ships that inch toward Seward and the invisible trinity of atoms of which we are speaking, this tripartite molecule that stretches and bends to the waves that rise from the earth to meet it.

Here in Alaska the seas are warming, expanding with warmth, their acidity is increasing as the acid, carbon dioxide, enters them. In the tiny village of Shishmaref, off the Seward Peninsula, rising sea levels have led to the town's evacuation. The permafrost in large sections of the Arctic is no longer frozen. The greenhouse gases—carbon dioxide, methane—that were once safely banked there, drift off into the atmosphere. And the Arctic sea ice thins and thins to the warming. In a blink of geologic time, it will all be gone.

❁

There was dense fog as we sailed toward the Hubbard Glacier. All you could see were bits of ice calved from its face. After waiting an hour for the fog to lift, the captain decided to turn back.

Titusville
and Tucson

We took our required course in Pennsylvania history in the ninth grade. I remember very little about it beyond the names Joshua Tittery, who was a seventeenth-century potter, and John Wanamaker, the Philadelphia merchant who more or less invented the department store. I remember that we discussed a small town that lies north of Pittsburgh called Titusville, where the first oil well was drilled in 1859. At the time, this hardly seemed like an epochal event to me, and I am sure I went home that afternoon looking forward to a game of touch football or some other activity of true importance.

Much later, I learned that the history of Titusville included a little-known visitor. His name was Dmitri Mendeleev, inventor of the modern periodic table of the elements. In 1876, he was sent by Czar Alexander II to gather information on the flourishing oil industry in the state of Pennsylvania. A record of his experiences in Titusville and elsewhere is given in a document recently acquired by the Chemical Heritage Foundation and still not translated from the original Russian. When I asked about Mendeleev's visit to her city, she told me that only one other person in the last five years had inquired about him.

Mendeleev's connection to oil and the oil industry is not well known, even among chemists. When I taught geochemistry, I would spend a few minutes discussing his abiotic theory of oil formation, which claimed that water deep in the earth might be reacting with iron carbides to form hydrocarbons, and that these, through time, might polymerize to generate the many organic compounds found in oil. This was an ingenious idea designed to explain the many mysterious black oozes that have been known for centuries throughout the world and that have been used for medicinal purposes, for lighting lamps, lubricating machinery, and even in the early art of mummification. Now, however, it is well known through the study of biomarkers that oil is a product of the transformation of once-living matter.

But when I think of Mendeleev, it is in terms of the periodic table and the great city of Saint Petersburg. In my sophomore year, that city and the elements came to be joined in my mind. I was taking a course in qualitative analysis and running through the infamous "qual scheme," with its blue lakes and flocculent precipitates and foul-smelling sulfides that eventually I came to like as though I were some kind of sediment-dwelling anerobe. The professor was particularly keen to point out how the reactions we were witnessing in the laboratory—the rain of calcium, strontium, and barium as white solids out of a pure clear liquid solution—made sense when you thought about their location in Mendeleev's table.

At the same time, in the gothic lecture halls of Pitt's Cathedral of Learning, I was enrolled in a literature course on tragedy. The professor, a distinguished Virginia poet by the name of Lawrence Lee, assigned the class some twenty novels, among them Camus's *The Stranger,* Stendahl's *The Red and the Black,* and an assortment of Russian classics, including *Crime and Punishment* and *Anna Karenina.* I had protested to my advisor that I had not wanted to take this requirement or any literature course. I was a chemistry major, after all, so why would I ever need it? Soon, though, I came to love the rich language and the images that were called to mind. *Anna Karenina* was begun only four years after the periodic law was announced—and when I thought of Mendeleev, I thought always of snow as it fell in St. Petersburg, while it fell just beyond the trolley windows where I read, on the streets of Pittsburgh.

Reihen	Gruppo I. — R'0	Gruppo II. — RO	Gruppo III. — R'0³	Gruppo IV. RH⁴ RO²	Gruppo V. RH³ R²0⁵	Gruppo VI. RH² RO³	Gruppo VII. RH R²0'	Gruppo VIII. — RO⁴
1	H=1							
2	Li=7	Be=9,4	B=11	C=12	N=14	O=16	F=19	
8	Na=23	Mg=24	Al=27,3	Si=28	P=31	S=32	Cl=35,5	
4	K=39	Ca=40	—=44	Ti=48	V=51	Cr=52	Mn=55	Fo=56, Co=59, Ni=59, Cu=63.
5	(Cu=63)	Zn=65	—=68	—=72	As=75	So=78	Br=80	
6	Rb=85	Sr=87	?Yt=88	Zr=90	Nb=94	Mo=96	—=100	Ru=104, Rh=104, Pd=106, Ag=108.
7	(Ag=108)	Cd=112	In=113	Sn=118	Sb=122	Te=125	J=127	
8	Cs=133	Ba=137	?Di=138	?Ce=140	—	—	—	— — — —
9	(—)	—	—	—	—	—	—	
10	—	—	?Er=178	?La=180	Ta=182	W=184	—	Os=195, Ir=197, Pt=198, Au=199.
11	(Au=199)	Hg=200	Tl=204	Pb=207	Bi=208	—	—	
12	—	—	—	Th=231	—	U=240	—	— — — —

Mendeleev's early version of the periodic table.

❦

After Lavoisier, there had been a great loosening up of chemistry. The science of matter moved away from its emphasis on the *forces* existing between atoms and focused instead on something that could actually be measured: *weight*. The English meteorologist John Dalton provided clear definitions of *atom* and *element* and described a method for determining the relative weights of atoms. Careful measurements followed and there were breakthroughs in the isolation of new elements, especially with the development of spectroscopy and electrolysis. By Mendeleyev's time there were sixty-three known elements and they displayed an array of properties that must have seemed to the casual observer stupefying at best. Some were gases, like oxygen and hydrogen; others were dense hard metals, like iron and nickel; still others, like lithium, floated like matchsticks on water. There was silvery mercury, a metal that pooled and skittered along the floor and whose very nature, to the Chinese emperors and the European alchemists, was a mystery. Some elements were highly reactive, like fluorine, which could etch glass and burn skin down to the bone, and others, like gold, were famous for their glittering inertness. There was, moreover, a long range of atomic weights that stretched from hydrogen, with a relative weight of 1, to uranium, whose atomic weight had been measured at 238.

How was it possible, in 1869, to make sense of all of this? Were there only sixty-three elements? Were there thousands more that no one knew about? Was the number of elements that made up the universe infinite? I have always found this the most vexing problem, the question of knowing whether there was any limit to the variety of particles that constitute the physical world or whether a little more probing might not just bring to light a snowstorm's abundance of unique atomic forms, numerous as the shapes of crystals that fall from the winter skies. Lavoisier, as he often did, summarized the situation as it was in his day: "All that can be said upon the nature and number of the elements is confined to discussions of an entirely metaphysical nature. The subject only furnishes us with indefinite problems." If the elements had given Lavoisier a headache, they must have given Mendeleev a migraine.

In West Lafayette, my friend Harv, with whom I was taking inorganic chemistry, would say, "It's all a bunch of memorization, a bunch of rules. Nothing ever gets explained in any deep, satisfying way until physical chemistry comes along." In part he was right, but I never thought of the periodic table in this way. To me it was a great plateau in chemistry's long ascent, a place that when you finally reached it you could see order in the bare landscape over which you had been struggling. It was so much more than just a chart on a wall, more than a thing to be sung out with mnemonics like the scales you had learned as a kid. I would tell Harv, "There's real beauty here. There's no way to overemphasize what Mendeleev gave us." And he would tell me, "If you want real beauty, real creativity, look to Schrödinger or to Pauli or Boltzmann." He scoffed at anything that had to be memorized.

But what impressed me about Mendeleev was the apparent simplicity of his approach. It was like Kepler seeing into the heart of the solar system, finding the rules that governed the planets, the elliptical paths from which so much that was baroque and unnecessary in the Copernican system just fell away. Or Newton condensing much of physics into the crafted parsimony of the *Principia*. Mendeleev arranged his elements in order of their atomic number, but instead of stretching them out in a long sequence from 1 to 238, he noted a tendency for repetition. For example, the element that follows fluorine in order of atomic weights is sodium (neon would not be discovered until 1898), and Mendeleev placed it in a column beneath lithium. The two have similar physical and chemical properties and are rightly placed in a common family. The next element, magnesium, shares characteristics with beryllium and so it was positioned below it in the table.

This use of themes and repetition may well have been inspired by the music of Robert Schumann. By his own account, Mendeleev and his wife and children occasionally gathered to play music at the family home. One evening they performed the Piano Quintet in E flat major composed by Schumann in 1842. The abrupt stops and the frequent repetitions associated with that piece were precisely the ethereal goad that Mendeleev needed to set down onto paper the rough sketch of his table. Hearing this reminded me of August Kekulé's famous dream of whirling snakes with tails in their mouths, a self-described dream from which came the ring structure of benzene.

A deep unfathomable creativity lies in the images that come to the prepared and waiting mind. Ideas, theories, and hypotheses in science derive from strange places: dreams and shifting ice flows, symphonies, the pure luck of a conversation in the park or in some stuffy office, Platonic musings on ideal worlds, chandeliers moving on an autumn breeze.

But beyond the image, as always in science, there must be tests, the hard stones of fact, proof, logic, and sometimes mockery and bitter debate. And it was like this for Mendeleev. But his genius lay not only in the conception but in the argument that would win him international fame even by the time he had arrived in Titusville.

His strategy, one that no one before him had employed, was to leave gaps in the table and to predict the existence of elements as yet undiscovered, and then to go even further and predict their properties. It has always sounded crazy to me, like walking off into thin air with the hope that some jutting rock will project out beneath your feet. But a great theory takes chances. If you had lined up the elements by atomic weight in Mendeleev's day, you would have found that beyond zinc (atomic weight = 65.37) lay arsenic (atomic weight = 74.92). Mendeleev could have chosen to place this in his table just below aluminum. But realizing that aluminum and arsenic were just too different in their properties to share a common family, he decided that there must be another element that would fit beneath aluminum, an element which would in many ways resemble it more closely. He gave this unknown entity a name: eka-aluminum. As far as anyone knew, it did not exist.

He went even further. He assigned eka-aluminum a set of properties. He predicted it would have an atomic weight of 68, a density of 5.9, and a low boiling point. He described the kind of compounds it should form with oxygen. The confirmation took a few years, but in 1875, Lecoq de Boisbaudran came upon a new element in the midst of a zinc deposit in the Pyrenees. He called it *gallium*, after the Latin name for France, and his careful measurements showed that it had an atomic weight of 69.9, a density of 5.93, and a melting point of 30.1 degrees Celsius. It was indeed the eka-aluminum that Mendeleev had predicted.

There were other successes along these lines. Mendeleev knew there should be, somewhere in the world, an element to fill in the blank space he had left beneath silicon. He called it eka-silicon, predicted its properties, including its detailed chemistry, and in 1887, Clemens Winckler, in Freiburg, discovered an element that had nearly the properties Mendeleev had foretold. He called it *germanium*. These and other predictions were convincing evidence that there was indeed an order to the building elements of the physical world, and that the fears of Lavoisier, that this misguided quest would result only in an "indefinite confusion," were happily far from the mark.

Exactly where all of these elements had come from, why there were so many when one—a single primordial hydrogen atom, round and small and perfectly made—would have easily sufficed, and what deeper underlying processes were at work to insure the structure of Mendeleev's table—these were questions whose answers would come much later, in our own time.

❀

The road west of Tucson takes you out past the Saguraro National Park, with its stately exotic cactus, toward tall mountains that in November are brown and barren. Against snow high above the road, the white domes of the telescopes emerged like a string of coral stretched across the highest ridge.

The ascent is steep and edged by deep valleys whose desert floors would take only one mistaken turn to arrive at with ease. There are switchbacks and scenic overlooks with signs that tell you the top is only five miles away. The road is just twelve miles long, but Wanda looked grim, with one of those expressions that says: "Remind me again, why are we doing this?"

The sign near the parking lot reads:

Welcome
Kitt Peak National Laboratory
Operated by the
Association of Universities
for
Research in Astronomy
Under Cooperative Agreement with
National Science Foundation

The names of the twenty-two participating universities follow.

By the visitor's center, there is a large, colorful mural done in bright oranges and blues and yellows, a representation of the heavens as envisioned by the Mayans. Wanda felt in her element then and may not have regretted the trip up there.

Kitt Peak Observatory is located on the Tohono O'odham Reservation, fifty-eight miles to the southwest of Tucson. At an elevation of roughly seven thousand feet, it rises high above the Sonoran Desert and is considered a sacred mountain by the Tohono people. To even build

141

this facility required delicate negotiations, which included a telescopic viewing of the moon by the tribal elders, who eventually gave their consent but placed constraints upon the use of their revered mountain. The facility includes twenty-four optical and two radio telescopes and the world's largest solar telescope. This last instrument was dedicated in 1962 by President John Kennedy, who called it "a source of pride to the nation."

I went with a small group to view this telescope and to hear about the kind of research that was being conducted here. The McMath-Pierce telescope is a huge piece of equipment. The main tower, on which the heliostat is located, is one hundred feet tall. The mirror atop this tower tracks the movement of the sun throughout the day and focuses a beam of sunlight down a two-hundred-foot shaft, which extends deep into the mountain itself. The light is eventually focused into a viewing lab beneath ground, where, through the use of spectroscopes, it can be used to examine the structure of sunspots, which may be thousands of miles in diameter, or to determine the spectra of various elements. The McMath-Pierce was the first telescope to detect the presence of water vapor in the sun and to discover isotopic helium.

In the viewing room, I was surprised to learn that seventy elements have been identified in the sun. Of course, hydrogen (91.2 percent) and helium (8.7 percent) account for most of the sun's mass, but there are also significant quantities of oxygen, carbon, nitrogen, sulfur, magnesium, and iron present, in addition to a host of lesser elements. The day I visited, the two scientists there were looking at the spectrum of helium in the sun's photosphere. The photosphere is the outermost region of the sun. It is where our light and heat come from. It is what sustains us, nurtures us, gives us warmth.

I have always thought it interesting that we found helium on the Sun before we found it on Earth. Using the newly invented spectroscope of Bunsen and Kirchhoff, Norman Lockyer found two absorption lines in 1868 that did not correspond to any known element. He knew he had a new element and he named it after the Greek sun god, Helios. Twenty-five years later the element was found on earth by Sir William

Two of the telescopes at the Kitt Peak National Observatory.

Ramsay, in the mineral cleveite. On earth, most of this rare inert gas is produced from the alpha-decay of uranium and thorium.

On the sacred mountain high above the desert, above the tall saguaro, among the white domed scopes of Kitt Peak, my thoughts turned naturally to the vast universe whose very nature was the subject of all that was there. All the negotiations, the hard sweat and labor of construction, the building of the sinuous road we had just ascended, the instruments of discovery—all of these are there to serve a single end. In this setting it was easy to think about large questions, to wonder about beginnings and ends, births and deaths, about origins and the way things are put together.

There are 70 elements in the sun; 114 elements are known to science, and all have been placed in the table. Mendeleev knew of only 63. Where all of these came from is a story that joins the chemistry of matter to the behavior of the distant stars.

✺

From the mountain, you look south toward Mexico, west toward California. Below is the desert. All around are the domes of the telescopes, the chute that guides the sun's light into the earth and the spectrographs. Everywhere there is matter piled upon matter, element upon element, the rock stability of the silicates anchoring the carbon spines of saguaro in the distance below. The fugacious atmosphere, rich with its gases, moves the labeled pines, and the chromium and cobalt of the Mayan mural catches the light of the sun. In the beginning, a few short minutes after the beginning, there were only two elements, hydrogen and helium, and these alone swirled and twisted in streamers and clouds through the newly minted universe. From these, everything that we see up here came and made possible the habitable and uninhabitable worlds, the ringed and jeweled and moon-filled worlds whose presence is a thing of wonder.

There is a story that has passed many tests, that shades into truth and has become truth, that all of this was born in the forges and furnaces of the stars, in the pulse of collapse and expansion, and in the unimaginable violence of explosion. It is this explosion that scatters the wealth of elements, as a beneficent hand might scatter gold coins.

The process by which new elements are made, brought into being in the dust storm of creation, is called *nucleosynthesis*. We know now that stars—like the out-rushing, fourteen-billion-year-old universe itself, the expanding red-shifted universe of Edwin's Hubble's observations on Mount Wilson—evolve.

A star forms from the tenuous clouds of hydrogen and helium that comprise a nebula. A nebular cloud is, in effect, the birthplace of stars and the locus of beauty to rival any landscape. The Eagle Nebula, with its three pillars, like the lonely tufas of Mono Lake, or the conch shells and rings of the Cat's Eye Nebula, are reminders of nature's boundless extent beyond the park boundaries and the scenic overlooks of our more common experience.

Through the Hubble, that masterpiece of engineering and craft, you can virtually see star birth occurring before your eyes. Here, dense regions of gas undergo collapse to form a rotating globule, and temperature and pressure increase. At the core of the rotating mass, temperatures soon reach the fusion point. At the periphery of the rotating disk, planets often form, as in our own solar system.

At the core of stars, temperatures are in the millions of degrees. The star's core is a place of relentless burning, the fires and light of the heavens. But the fuel is not the ordinary oils of Mendeleev or the coke of steel. It is the simplest of all elements, a single proton endlessly circled by an electron, or, as we now know, by a tiny cloud, denser here, thinner there, a patchwork of probabilities. At the temperatures of the core, hydrogen can undergo fusion and become, through a chain of reactions, the element helium, which has two protons and two neutrons in its nucleus. It is this hydrogen burning that keeps the sun from collapse. Always, in any star, it is the balance of these two, the outward pressure from the burning and the inward force of gravity, that maintains a kind of equilibrium. Our own sun, which is five billion years old, has enough hydrogen to maintain this balance for another five billion years.

Most of a star's lifetime—an estimated 99 percent—is spent in the conversion of hydrogen to helium at about ten million degrees Kelvin. But as the star evolves, higher elements are created in an environment no Cotton Mather of hell and brimstone fame could have ever envisioned. The making of higher elements requires more extreme temperatures because fusion involves bringing together nuclei of ever greater positive charge. When hydrogen at the core is finally exhausted, converted into the invisible ash of helium, the star begins to collapse under its own gravitational mass. This process generates enormous heat and enormous nuclear velocities, which are sufficient to overcome the electrostatic repulsion and to allow for the fusion of helium into carbon. This process requires about one hundred million degrees. Heavier elements—oxygen, for example—are created in a similar way, but at still higher temperatures.

The nuclear reactions needed to build the bricks and mortar of the material world up to uranium are far too numerous to list here. But it is interesting to note that nickel-56 is the heaviest element generated by the addition of a helium nucleus (alpha particle). This unstable form of nickel then eventually decays into iron, an element that is especially abundant in our solar system. On earth, it is the element of building and progress, of the ribbed steel and girders of New York and Chicago and Pittsburgh, stacked toward the sky.

Elements more massive than iron often form by the slow process of neutron capture (s-process) or by the more rapid capture process (r-process) frequently associated with a supernova. When the core of a star contains mostly iron and can no longer undergo significant fusion, the star collapses. It is estimated that such a collapse takes less than a second. The resulting explosion sends atoms in a great mist across the universe, and these atoms, in a generous act of cosmic recycling, become part of the nebular cloud from which new stars are born. Our sun is thought to be a third-generation star formed, in part, from the elemental debris of earlier starbursts. Some of the heavier elements from which our solar system was formed came from this violence of long ago.

Toward the end of our day, we stopped again at the visitor's center. For sale at the counter was a small telescope, a replica of the one Galileo had used for his book *The Starry Messenger.*

Nearby, there was a framed quote from Johannes Kepler. It read:

We do not know for what purpose the birds do sing, for their song is pleasure since they were created for singing. Similarly, we ought not ask why the human mind troubles to fathom the secrets of the heavens. The diversity of the phenomena of Nature is so great and the treasures hidden in the heavens so rich, precisely in order that the human mind shall never be lacking in fresh nourishment.

At Kitt Peak, who could help but agree. For me, it was a pleasure to see the two great astronomers, who had never met in life, side by side, honored here together on the mountain.

※

Shortly after Tucson, Wanda and I were in Titusville, Pennsylvania. We stayed for only a day. We walked around the grounds, explored the well that Colonel Edwin Drake had drilled. His crew struck oil in 1859 at a depth of 69.5 feet. There was great celebration at this find—the first oil well in the world. Titusville became a boomtown overnight. Wanda mentioned that the smell of oil was still in the air. There was even a faint taste of it on my tongue.

In the park, there were drilling rigs from different periods, showing the evolution of the industry. Huge flywheels, "dunking birds," derricks, cast-iron boilers, spindly rods connected to pipes. There were photos of the men, long gone, who had worked the fields. The drillers and tool pressers, roughnecks and roustabouts—the standard crew. Amid the machines that had transformed the world, had changed for centuries to come the thin chemistry of the atmosphere, there was cropped green grass and broad-leafed rhododendron and black oak. The leaves of mid-November had all fallen. Wanda said, "It looks like a sculpture garden."

Though there was no mention on any of the signs of Dmitri Mendeleev, you could imagine how the place looked when he came here. In time, it would be learned what oil contained. Mostly it was carbon and hydrogen and some oxygen arranged in thousands of compounds, but there were heavy metals, too, and many of them had been precipitated by sulfides from the waters of ancient seas.

It was a pleasant, overcast November day, and we walked with our coats opened. Before we left, I took a picture of Oil Creek. It was broad, smooth, and dark. Across its waters, a mountain of star-fused elements rose straight up from the shore.

Vienna

At the Hotel Schweizerhof, no one had ever heard of Boltzmann or his tomb, but everyone knew about the Central Cemetery, where the prosperous and celebrated of the city lay.

When I took the tram there and had walked for twenty minutes from the main gate to the musician's corner, I sketched the location of the graves, which included those of Beethoven, Franz Schubert, Mozart, Johann Strauss, Brahms, and Arnold Schoenberg. People with cameras were bent over the tombs, reading inscriptions, then standing back to snap a photo. There was silence or an occasional whisper. It was a Sunday morning in May.

Just across the road was the tomb I had come to see. The sunlight was very bright and there were tiny flowers in the green grass. The family names were all there on the stone, including his wife, Henriette, who had outlived him by more than thirty years. Above the raised visage of Boltzmann, near the very top of the monument, was the famous equation:

$$S = k \log W$$

Whatever muted sounds there had been from across the road were gone now. I was alone standing in front of Boltzmann's tomb. For years I had thought about this place, and then for decades it had disappeared, the memory of Edgell and Boltzmann had been lost among more pressing concerns. But now on a sunny day in Vienna everything had returned: the morning down at Curleys with Buckley and Harv and Rita

149

and Rilke's words about an "exhausted nature" calling back her lovers; and that year in Blacksburg, the year of randomness, like one of Boltzmann's molecules, crossing zigzag into unexpected spaces beyond any plan or narrative I had ever mapped for myself, the vast white continent whose name I had never spoken spread before me that night in the trailer when Benoit's offer came, the day Field said it would be fine to go, and all that followed, the plans and packing, the choppers making their way through heavy storm and whiteness, the papers written and published, the conferences and lectures, and my daughter and Joseph in the rugged Asgards signaling with their mirror that all was well.

I realized at the tomb that the European trip had really been about this, about standing here, about sensing in the randomness and tangle and brokenness of the world its improbable beauty.

What was yet to come, I had no idea: the illnesses, unexpected, came some years later from some hidden place, somehow foreshadowed in the law inscribed on his tomb in stone, not more than ten feet away from me. Entropy is about death, Harv used to say.

❧

Harv had been right. It was better to think of entropy in terms of probabilities rather than heat engines, no matter how ideal those heat engines might be. It was better to think like Boltzmann did, in terms of molecules, of caged birds trying to escape confinement, of wolves set free onto a thousand square miles of tundra.

Sometimes I would ask my classes to envision a blue gas in a small vial set off in the corner of the room. What would happen if I opened the glass stopcock? Would the gas just stay there in its precious untroubled seclusion? Or would something else occur, something entirely disruptive? Everyone knew how the world worked; they knew what made a gas, the flights and energies of its molecules, and they would tell you, almost in a single voice, that the gas would disperse, would spread out into the room. That perhaps it would cast a shade of faint blueness on the very air that they breathed.

*Stamp issued by the Austrian Government
to honor Ludwig Boltzmann.*

And if the door were opened, it would spread into the hallways, to the ceilings, down the stairs, and out into the air. Gradually it would encircle the Earth.

Suppose the classroom door had not been opened; would the gas ever return to its vial in the corner of the room? Most of the students would say *impossible*. And I ask, "What if you waited a thousand years. A million? A billion?" A girl in the back of the room says, "Well, I think there may be a small probability that one day, who knows how long from now, all of the molecules will find their way back."

The magic word: *probability*. There is a finite probability that the molecules will one day return, that the vial will grow more intensely blue. Boltzmann would have answered this way, would have included the idea of probability, however small in this case, in the laws of physics.

The second law of thermodynamics, the one inscribed in stone before me, is a probabilistic law. It says that the entropy, S, is related to the number of ways in which a particular state can be realized. Entropy, in effect, is a measure of disorder. The blue gas released into the room has a higher entropy than the same gas confined in the tiny vial. Released into the hallways, it has an even higher entropy, a higher level of disorder. Entropy is a measure of molecular chaos, of scope and freedom, of "spread-outness." And the world, as it turns out, moves always, spontaneously, in the direction of increasing entropy.

Humpty Dumpty's tumble from the wall ends in the symmetrical confinement of the egg being broken and replaced by scattered shards and spread yoke, an albuminous ruin, which could, if all the countless molecules moved in some improbable concert in just the right way, reassemble into eggy perfection as though nothing had happened. But in all the unwritten history of the universe to come, this reassemblage is unlikely.

The entropy law points out the direction of time. Arthur Eddington called it "time's arrow." The coal burned in the furnace, the glowing ember turned to ash cannot be recovered from the gases and oxides that remain. It is gone. And the oil of Titusville, consumed in the homes and factories of Pennsylvania, will never again become the ancient mixture of hydrocarbons that so fascinated Mendeleev. Nor will the sunlight gone down the long shaft of the McMath-Pierce telescope and into the spectroscopes and computers that analyzed it ever return to the Sun. Harv, in his lugubrious way, under the dark skies of Indiana, had said entropy was all about death, and in a way it was.

In "West Running Brook," Robert Frost said that "the sun runs down in sending up the brook," and indeed it does: the hydrogen turned to helium, its fusion energy gone to some far-off place, to some Jersey shore in summer, to some autumn brook, to the light that finally breaks the chlorine free. This energy runs one way as does the sun itself. "Entropy! / Thou seal on extinction / Thou curse on Creation," John Updike wrote. "All change distributes energy / Spills what cannot be gathered again." The molecules and nuclei by which we power the world are the grapes of the vineyard, the wheat of the fields, here for the moment and then gone, their content transformed.

❦

Molecules are so real to us now that they have become as common as tableware, even though no one has ever seen a single molecule or atom in anything but the fuzziest photographic likeness. That the ontology of these entities was ever a subject of debate—and so recently as the late nineteenth and early twentieth centuries—seems inconceivable. Yet, try as he did, in debate after debate, Boltzmann could not convince his brilliant colleagues Ernst Mach and Friedrich Wilhelm Ostwald (one of the founders of physical chemistry), among many others, that these tiny particles, despite their ghostly intangibility, really existed. It was odd, as Rita had remarked down at Curley's place after her first year of chemistry, that people, so recently, just didn't get it.

In chemistry, it was so much easier to think of processes in exactly the molecular terms Boltzmann had laid out—as models and visions rather than equations alone. When Paul Field and I had finished our study of the molten salts, we were surprised to see that the gases we dissolved in liquid sodium nitrate actually did so with a far less negative entropy than the models had predicted. The entropy was negative, yes. But much less negative. How could that be? We were taking a relatively unconfined gas and placing some of it in the confined space of a hot ionic nitrate melt. So, of course, the entropy of solution would be negative. But the fact that it was only slightly negative meant that somewhere some disorder was being introduced. Somewhere in the system, which of course included more than the dissolving gas, little bursts of chaos were breaking out.

Field and I began to talk about this. We were now entering my favorite realm, the one beyond the building and testing of equipment, the glassblowing and T-joints, the tedious collection of data. Anyone coming into the lab would have seen the two of us sitting there sipping coffee, joking at one moment, racking our brains the next, and occasionally falling into a meditative silence.

But we were not really in that room. We were only apparitions. In truth, we had long departed. We had joined the sodium ions, the nitrate ions, and the atoms of argon gas. We were among these and of the same dimensions as they were, briskly walking, colliding in our quickened pace in search of the rarer argons that seemed lost among the ions.

We began to speak to one another from this Lilliputian world where we were hardly unusual, each of us about the radius of an atom, no more than that. And our reports were surprisingly similar. The chargeless spheres of argon, we agreed, had tunneled their way in among the ions, filling for a billionth of a second the interstitial spaces here and there. Down in the melt we inhabited we could see that. We could see, too, that for an instant, the argons had come between the charged ions, come between the attractions of positive and negative, and had sent them skittering off here and there, flying in momentary chaos and randomness. That was it! Although the gas was more confined than it had been before dissolving, had lost entropy, the melt had gained it, had become more entropic. On balance, the entropy of the process as a whole was still negative, but surprisingly less than anticipated.

When we emerged from our sojourn there was sunshine streaming in through the morning windows and our coffee had grown cold. We wrote the whole thing down right away, like awakened dreamers writing of quickly dispersing dreams.

I had read of another case where the entropy of solution came as a big surprise. It involved dissolving the rare gases—helium, neon, argon, and so on—in water. Here, a certain negative entropy was expected, but the results were wildly more negative than anyone had predicted. In this case the water reacted in such a way as to perform a kind of architectural shift. As if to exclude the gas entirely from its bulk, water built little structures around the invading molecules, producing patches of hydrogen-bonded order that lowered the entropy to a truly anomalous extent. This idea was first developed by Henry S. Frank and has since been shown, through spectroscopic and other evidence, to be true.

When I came to know him personally, Henry Frank often spoke of the long history of our efforts to understand water. It began with Thales of Miletus and extended down to the present time. Water was the most important liquid on the planet and possibly in the entire universe, and despite centuries of research it was still mysterious. Henry Frank always lamented that he would not be around to see how it all turned out.

❦

The second law of thermodynamics makes the radical claim that in every spontaneous natural process, the entropy of the universe increases, its level of disorder increases. Sometimes it does not seem this way. When the snowflake forms, crystallizes in its six-fold beauty, it seems as though order in the universe has become, to a tiny degree, greater. A snowstorm, with its countless flakes drifting through the gray December sky, produces order upon sweeping order. But think of this: Each time a tiny flake coalesces around a particle of dust, each time the molecules of water vapor congregate in their fibrous linen branches, chemical bonds are formed, and each bond releases in its infinitesimal creation a firefly burst of energy into the sky, where waiting nitrogen, oxygen, and the rarer argon and carbon dioxide are there to absorb it and, in so doing, to dance to the new energy in increasing agitation. When the snowflake's surroundings are taken in account, the disorder of the universe has indeed increased.

Back in West Lafayette, one of the best lecturers at the university would demonstrate this point. As he gave his talk, he would walk slowly about, a paripatetic in the Greek tradition, with a few incongruous balloons floating above his head. As he walked, he took out a lighter and casually set the long strings, dangling near the floor, on fire. The fires rose and reached the balloons, which he had filled with hydrogen and oxygen gas. One by one the balloons burst loudly into flame.

The gases had combined to form water: one molecule of oxygen plus two molecules of hydrogen produce two molecules of water. He asked next whether the reaction, as he had written it on the board, had a positive or negative entropy. The students thought about this for a few minutes and one said, "It's negative, because three molecules of reactant have given two molecules of product."

"Excellent," the professor said. "Now would you call that a spontaneous reaction?"

A student raised her hand and said, "Sure, that was absolutely spontaneous."

"But," the professor said, with a *gotcha* smile on his face, "the entropy law says that all spontaneous process must occur with a positive entropy. The entropy of the universe must increase."

And, indeed, he had them. The class of well over a hundred students sat in near-silence, looking down at their notebooks, rubbing their chins, fidgeting and whispering to one another. Had they just witnessed the breach of a sacred law? Apparently so.

Then there was a barely audible voice from down front. The girl was saying, "But doesn't the law always include the surroundings, not just the three molecules we have talked about? I mean everything?"

A smile lit Professor Davis's face and he bobbed his head up and down.

"I think you've got it," he said. Then he explained how the reaction had generated lots of heat, and the heat had caused the atmosphere around what had once been a balloon to expand and become energized. It had ripped open the balloon itself, warmed the rubber that contained the gas, and, generally, produced all kinds of local disorder. When you added the positive entropy change undergone by the *surroundings* to the negative entropy change associated with the *system* (the reacting gases), there was no question that the net entropy of the *universe* had increased.

It is hard to keep the surroundings in mind. When we build a city, we are focused on the orderly structures: the roads, the neatly placed parks with their greenery and benches, the inscribed statues of our heroes. Yet we forget the vast chaos we have created in their construction: the dark tunnels of coal we have dug out, the steel we have taken from burning coke and iron, and all the metal casting and slag heaps that have lit the night sky and the oxides of carbon traveled far on the winds of the earth. We know, on the balance sheets, that our little pockets of order have come at some expense.

❦

When Don Canfield and I first started work on Lake Vanda, what interested us was the way in which metals were removed. Vanda, like most of the lakes in the Dry Valleys, was a closed-basin system. It had, in this case, a single inflow, and absolutely no outflow. There were no streams exiting the lake as there are in most of the lakes we are familiar with. The water balance was maintained by an exotic mechanism, in which a certain amount of ice was ablated from the ice surface each year in the dark, sand-blasted winter. The amount of ablation had been measured by the New Zealand hydrologists, and it pretty nearly equaled what was entering the lake from the Onyx River during those six or eight weeks of flow. During the early days of the study, we just weren't certain what we would find, and at times we expected that elements like copper and lead might be at such high concentrations that they would act as poisons to much of the biota. After all, the ordinary ions that had been studied—calcium and magnesium, for example—were all at much higher concentrations than any of the more familiar lakes like Walden Pond, Lake Mendota, or the Great Lakes of North America.

But when we first began our research, we encountered problems we had not anticipated. As we drew our samples from deeper and deeper in the water column, we were struck by just how saline the bottom waters were. Although the upper waters, just below the thick ice cover, were fresh enough to drink—and, indeed, I had enjoyed many a delicious cup of cold Vanda water, which was purer and clearer than the freshest spring—the bottom waters, down at a depth of two hundred feet or more, were seven times saltier than seawater. Just a sip of this liquor was enough to give a bad taste for an hour or more. Why was it so salty? And why was the composition, mostly calcium and chloride ions, so unusual? How did a calcium chloride brine ever form in this lake, especially when the Onyx River was composed mostly of sodium and bicarbonate ions?

An extremely high concentration of anything in one place is always improbable. The blue gas in the small vial, when given the chance, quickly dispersed throughout our classroom. The same should be true of the ions near the bottom of Lake Vanda. And indeed they were diffusing, doing those little hop-skip motions that ions do, colliding with water molecules, all the time relentlessly spreading upward as the second law of thermodynamics requires. They were moving from a less probable to a more probable state, from order to disorder.

But this upward march was slow. We could calculate, based on how the concentration of chloride changed with depth, that these ions had been diffusing for twelve hundred years. It was a number very close to the one Alex Wilson had gotten, using his data from back in the sixties. But where had this deep pool of calcium chloride come from in the first place? Maybe the Onyx River, under evaporation, had contributed some of it, but certainly not that much.

To confirm this, I called Hans Eugster at Johns Hopkins University. Eugster had spent many years working on the salt lakes of the African Rift Valley. I asked him whether, in his experience, he had ever found a calcium chloride brine that had derived from sodium bicarbonate stream. "Impossible," he said. "Doesn't happen." We chatted for a while, but I had my calculations and instincts confirmed.

Canfield and I worked on this problem for a year and determined that the deep brine must have come largely from below the lake, from an ancient groundwater that had at one time seeped into the bottom of the basin. The brine, we concluded, was a mixture of ions from the evaporating Onyx and a mysterious source brewed over time in the deep bedrock. We published our results in the best international journal of geochemistry and left it at that. Not long afterward, a research group from West Lafayette, Indiana, wrote that they had studied chlorine isotopes from the Vanda brine and had concluded that our two-source hypothesis was, indeed, correct.

Following Wilson and our own data set, we discussed in our paper how the diffusion curve for chloride and the long upward twelve-hundred-year drift of ions must be telling us something about climate change: Twelve hundred years ago, a great flood of fresh water from the Onyx River, from the melting glaciers onto the Vanda brine, signaled a rapid warming in the Dry Valleys. This was long before Scott, or any human being, had ever seen the Antarctic continent. But coincidence or not, the warming came at nearly the same time as in Europe,

La Fin du Monde. A woodcut by the artist and astronomer Nicholas Camille Flammarion showing a version of the heat death.

the beginning of the Medieval Warm Period in AD 800.

The spreading out of ions in a distant lake was yet another demonstration of the equation inscribed on Boltzmann's tomb.

❋

When Harv talked about entropy and death, he was usually referring to himself or his parents or friends. He was taking about illness and things that he feared. But in the late nineteenthth or early twentieth centuries, people thought about how the second law of thermodynamics might be ringing the golden bell for all of us.

There is an illustration that appears in the pages of *La Fin du Monde* by Nicolas Camille Flammarion. It is entitled *La miserable race humaine perira par le froid*—the pitiful human race will perish from the cold. It shows what appears to be a family or an extended family on an ice-covered lake. A woman in animal skins hugs a child to her; an old man lies sprawled on the ice; a tall, thin man, without shoes, stands upright clutching his chest; and there is a figure crouching in the distance. Behind them are snow- and ice-pinnacled mountains. There are no fields or food anywhere in sight, and the distress is everywhere palpable.

Every log and lump of coal had been burned to ashes. The sun itself, in the cosmology of the day, had run its course. There was no conceivable form of warmth, no escape from the "heat death." The sand in the hourglass lies dead at the bottom.

But this is an unlikely ending. Gradually the Sun is becoming more luminous, hotter, so that in scarcely more than a billion years it will be hot enough to drive into space the water vapor in the Earth's atmosphere. In three and a half billion years, its hot-plate temperatures will be sufficient to sever the hydrogen bonds of water and to turn the oceans into boiling mist. The snows of Kilimanjaro, the ice of the Mendenhall, the vast East Antarctic Ice Sheet will be long-distant memories, coiled somewhere in some dead matter, hidden in the hidden rocks. Eventually the sun will become a "red giant" and engulf the inner planets in its redness, before it collapses into a "white dwarf" and cools to the ambience of the universe. Today, in the time of telescopes and cosmology, we know our future, and it is a future of fire, not of ice.

Cambridge

In the Dry Valleys, we were miners, but we cut through no stone, left no tailings mounded on the banks of rivers, burrowed and cut through no mountains. We were fishers without rods, reels, or feathery lures. Our ice holes opened onto an ancient silence, onto water that was blue, crystalline, and still, even through wind-blasted Aprils and Augusts.

In springtime, our metals came in small streams formed in the chutes of massive glaciers. The river seized from soils, in the cupped buckets of its passage, a trace of copper, a trace of nickel, a hint of cadmium, and brought them in intimate tangles of water in a gleeful sound-filled rolling to the lake. The same metals, the same atoms that in other lands were dug in abundance from the dolorous mines, were so rare in the waters of the Dry Valleys as to be seen with only the right quanta, the right wavelength, piercing into the tiny clouds in which they hung for an instant. The science of spectroscopy made possible their measurement and measure. *That* science began in the Cavendish Laboratory, on a narrow street in Cambridge, England.

Sometimes there are small villages, maybe only a few square miles in size, that somehow become the centers of impossible genius. Concord, Massachusetts, is one example. Not far from the town center is the pond which Thoreau made famous in his classic, *Walden*, which has become the founding work of nature writing and environmentalism in America. Only streets away from the town square are the houses of Ralph Waldo Emerson, Nathaniel Hawthorne, and Louisa May Alcott; Herman Melville, who lived to the west in the Berkshires, visited frequently. Collectively, this group gave birth to the so-called American Renaissance in literature and generated a definitive body of work, which included novels, essays, and poetry—all of which are still widely read today.

If you had to choose a town of similar dimensions whose citizens had an incomparably greater revolutionary impact on the way we see the world, it would be Cambridge, England. For it was here that Newton did much of his work, as did J.J. Thomson, Ernest Rutherford, James Watson, Francis Crick, and a host of others. Kuhnian paradigms, over the centuries, blossomed here as in a visionary's garden.

J.J. Thomson intrigued me the most, not just because of his youth and ingenuity and the seeming high regard in which his students held him, but because of what he had created. Who could think of atoms today without his singular creation—*electron*s? Or chemistry, itself, with its single, double, and triple bonds, like hawsers and ships' lines holding the world together. What had I been studying if not these tiny swats of matter, moving in a concert of unseen silence, in puffy clouds of what might be or what surely was the connective tissue in all that we see?

In Indiana, Walter Edgell had begun his story of matter with the way Thomson had pondered the eerie glow of the big-bellied Crookes tubes. Thomson concluded that something was *flowing*, a ghostly river of strangeness, within the tube, between cathode and anode. One of his crucial experiments was to run the cathode beam through an electric field. This had been tried by other investigators, among them Hertz, but Thomson was able to create the right conditions, the right gas pressures and voltages, to achieve the important result. What he observed in the thin glow of the tube was an upward curvature of the beam toward the positively charged plate. Clearly this beam was not simply light, but something else, something negatively charged and drawn to a positive

Sir Isaac Newton's Mathematical Bridge in Cambridge, England.

electrode. Moreover, it didn't matter what metal was used as the cathode (iron, nickel, copper, zinc); all gave rise to the same negatively charged stream.

Thomson experimented with cathode beams in magnetic fields and found that their behavior in this setting was also consistent with a negative charge. Later he developed an apparatus that combined electric and magnetic fields, both of known strength. Using this arrangement, he determined the charge to mass *ratio* of the electron. In Chicago, Millikan's brilliant oil-drop experiment allowed the electron charge to be measured independently. By 1909, the electron had become a well-characterized particle, a whiff of charge, a tiny fraction (1/1836) the mass charge of the hydrogen atom.

Edgell seemed fascinated by the way in which Thomson, Rutherford, and Bohr had puzzled over the placement of this nearly weightless wraith in the body of the atom. Thomson imagined that there must be positive atomic charges to counterbalance the electrons. He envisioned a curious confection called the "plum pudding model," in which the plumy electrons are embedded in a positively charged "pudding." In Thomson's words, as they appeared in the journal *Philosophical Magazine,* "atoms of the elements consist of a number of electrified corpuscles enclosed in a sphere of uniform positive electrification."

Atomic theories in the early decades of the twentieth century had short half-lives, as Edgell smilingly pointed out, and Thomson's model was no exception. In 1909, his former student Ernest Rutherford was studying the interaction between alpha particles (helium nuclei) and ordinary matter. In one of the more elegant experiments in the history of physics, he shot a beam of these particles at a thin sheet of gold foil. The expectation was that the positively charged alpha particles would be bent slightly by the positive particles within the gold. But on that occasion something strange happened. Some of the alphas actually rebounded, came rushing back from the gold foil, and hit the source— like a baseball ricocheting from a cement wall. Rutherford was stunned. In a letter to a friend, he said:

> It was quite the most incredible event that has ever happened to me in my life. It was almost as incredible as if you fired a fifteen-inch shell at a piece of tissue paper and it came back and hit you.

What was causing this flimsy sheet of gold foil to withstand the mighty rush of a massive alpha particle, and moreover, to turn it around and send it back from where it had come? Rutherford thought he knew, and his calculations bore him out. The particles could have been repelled with such force, he reasoned, only if all the positive charge in the gold atoms were compressed into a hard, dense, compacted point of matter—into a nucleus. But where were the electrons? Rutherford argued, and his brilliant student, Neils Bohr, showed mathematically, that the electrons must be orbiting this nuclear sun like so many tiny Copernican planets, so many "Galilean stars." Suddenly, with one crucial experiment, with one shuddering insight, it became clear that atoms were not solid little billiard balls, or plum puddings, but mostly emptiness, void—a tight nucleus with its distant turning moons.

There was now a cosmic loneliness at the heart of every solid object. You may as well have been an astronomer at Kitt Peak, contemplating the cold distances between planets, between stars, between receding galaxies; or Edwin Hubble on Mount Wilson contemplating the great out-rushing with its red shift, its fourteen billion years of time teased from those sightings. Even solid Pluto was empty to its core.

Of course, the physics of Rutherford was only the beginning. Bohr saw the possible arrangements of the solitary electron in hydrogen, how it could jump from quantized orbit to quantized orbit, the vaporous blue light giving rise to the four brilliant Balmer lines when passed through a prism: red, green, blue-green, and violet. Finally, matter and light were linked.

Heisenberg, Born, Schrödinger, Dirac, and Jordan—the magicians of the "Quantum Age"—had yet to capture the electron as wave, as mere mist of charge, only its probable location known. They had yet to give deepened logic to the elements, to the play and recurrence of their themes: the nobility of platinum and gold, the vigor of metallic potassium and cesium, the vivacity of fluorine and chlorine.

But after two short years, from 1925 to 1927, it was understood at last why hydrogen behaved as it did, its reactivity as the diatomic element: the sad image of the Hindenburg at Lakehurst, collapsing in ruins. The same with oxygen: Now you could rationalize its valence, its propensity for metals, its gradual evolution over the life of the planet to its present level of 21 percent. The red and orange rusts follow exactly from the electron structure of iron and oxygen.

Mendeleev did not live long enough to know why the elements could be arranged as they were. He never knew the theoretical underpinnings of his discovery.

Along the banks of the Cam and not far from the fabled Cavendish, I am no doubt walking in the footsteps of my first mentor,

Alan Clifford. Clifford often talked of his year in Cambridge, working with the deadly reactive fluorine gas to synthesize some exotic compound, perhaps even xenon hexafluoride, which was finally made some years later at Argonne National Labs. There was general rejoicing at this preparation, since it was the first time a rare gas had ever been coaxed into union with another element.

Success or not, Clifford seemed a man of the world. Dining with him and his research group at a local restaurant in West Lafayette, I realized that he was the first person I had ever seen eat with chopsticks. At that dinner, I heard for the first time the word "punt," in its English sense: flat-bottom boat, long pole, shallow river, the peace and magic of water as it makes its way slowly among the grasses and lawns.

Along the Cam there is a bridge that runs from a lovely tree-lined bank directly into the stone buildings of Claire College. It must be only a hundred feet or so across the river at this point. The bridge rises about twelve feet above the water at its apex, and is a beautiful and simple construction that is said to have been built by Sir Isaac Newton himself, with the wood fitted together by him without the use of nails.

It was just before noon and I was looking down at two long, flat boats as they slowly approached me. The punts, each of which contained a young couple, moved unsteadily along, swerving drunkenly left and then right, as the standing punters with their long poles tried to rectify their courses. Eventually there was a collision, one boat ramming into the other at a perfect right angle, as though the spirit of Newton were directing events. The collision was unintentional and there was wild laughter. Miraculously, both punters—hopping around on one foot and looking ridiculous—were left standing. Since the boats were now drawing closer to the bridge, I could make out the faces of their occupants. One of the punters, despite his good nature, appeared clearly frustrated and decided to launch off in a new direction that would take him out of the tangle he was now in. Straightening up and setting his jaw, he shoved the long pole squarely into the water, found the bottom of the Cam, and gave a mighty push.

No sooner had the resolute punter lodged his pole in the sediments of the Cam, than his punt shot across the water, with the sediments applying to him as much force as he had applied to them. His light craft slammed into the ancient walls of Claire College, and with that collision the punt shot back, again in the opposite direction, and struck the second boat with such force as to send all four punters tumbling into the water.

Amid laughter and much splashing and paddling was a "Newtonian moment," a perfect demonstration of the third law of the *Principia*—to every action there is always opposed an equal reaction.

※

In the past five centuries, the kaleidoscope has turned in its blue-green-red-yellow mix and somehow it has all become richer, more nuanced, more splashed with time and hiddenness. Thomas Kuhn tells us that after a scientific revolution, we live in a different world. It could be a world that is simply more open, far vaster than we had ever imagined; or a world that is beyond our comprehension so that in the end we are forced to throw up our hands and exclaim, "Well, that's just the way it is"; or, again, it could be a world that lies far below our senses, that we know is there, that is the vibrating linked presence of even the most common objects—tree and table and rusted iron—of our everyday experience; it could be the shifting earth on which we stand, that we can no longer take for granted as a solid, immovable platform, an earth of great improbable age from which life has sprung and ramified into form over form beyond counting, like the glittering galaxies floating in the eye of a Hubble.

Some revolutions seem to burrow more deeply than others, seem to cut to the existential bone. The "Little Monk" in Bertolt Brecht's *Galileo* wonders how the new revolutionary science of Copernicus, Kepler, and Galileo would be received by his parents, who are "peasants of the campagna, who know about the cultivation of the olive and not much about anything else." "They have been told," he says, "that God relies upon them and that the pageant of the world has been written

167

around them. . . . How could they take it if I told them that they were on a lump of stone ceaselessly spinning in empty space, circling around a second-rate star? What then would be the use of their patience, their acceptance, their misery?"

And so we argue about Darwin and evolution, even now, in the same misguided way, despite all the evidence. We build museums to show that the world is a mere six thousand years old. Once, in a lecture, when I spoke of the great age of Lake Vanda, of the glacier-carved valleys themselves, I was interrupted by someone in the audience and reminded that the world itself was only a few thousand years old. The Creation Museum, not far from where I live, so professional in its design, displays a flood-formed planet where dinosaurs and human beings trod the same leafy Edenic space. On any day you can find in the parking lot of that museum the license plates of cars that have come from thirty or forty states.

In my lab in Boyd Hall there was always great excitement when the water samples arrived from McMurdo. We had carefully packed them in Nalgene bottles—one bottle from each depth in the lake—and the samples had been acidified in the field, usually minutes after collection, by adding a few milliliters of ultrapure nitric acid. We wanted the pH to be two or less to insure that no metals adsorbed to the surface of the bottle.

As we opened the wooden crates the wood squeaked and the heavy lid fell off and struck the floor with a thud. We scrambled through the box to make certain that none of the bottles had broken or leaked, and when we found that they hadn't we applauded fate and UPS and began setting them up in long neat rows along the lab bench.

Within a few days, we brought out the organic solvent, the buffer solution, and the chelating agent and prepared to extract the metals in an effort to concentrate them in a small volume. Every metal that we studied had such low concentrations in the lakes and streams of the Dry

Valleys that this was the only strategy that would allow us to measure them, even using the extremely sensitive instrument we had sitting on the next bench, the graphite furnace atomic absorption spectrometer, or GFAAS for short.

The tiny graphite tube of the atomic absorption instrument smoked and flashed, and the lamp sent its beam through the cloud. The absorption was a signature that set apart nickel from copper from lead and zinc and cadmium. With our instrument we measured all of these and many more in the Antarctic streams—waters purer than any Walden—at the very end of the earth. As the numbers came in and were printed out on thin scrolls, we were standing astride three continents, and at the end of hundreds of years of chemical and physical history that made it all possible.

*

The building in Free School Lane had been a shrine of sorts, to which so much of modern physics and chemistry could trace its origins. To see it now in its present state, the housing for a quaint collection of university departments, was like seeing a body drained of life, a mere enclosure for what had once been. In 1972, the physics department had abandoned the Cavendish for more spacious quarters not far away. Inside, the Maxwell Lecture Hall remains, and the freshmen of Cambridge still learn their physics here. On the wall facing Free School Lane is a plaque that reminds the odd passerby of the revolutions that had begun inside; where twenty-six nobelists, including J. J. Thomson, had imagined, experimented, refined, calculated, and published their way to a new reality.

*

When Barbara and I taught our course on Paradigms and Revolutions, we usually began with readings from Karl Popper and Jacob Brownowski. We wanted our students to think about the nature of science and to consider that it might even be *creative*, not unlike art and poetry. Most of them had never thought about these matters. For them, science was a bunch of facts and equations to be memorized, problems to be solved; the old plug-and-chug routines of my own education.

After discussing selections of Popper's *Conjectures and Refutations* and Bronowski's *Science and Human Values,* we read James Watson's *The Double Helix.* We were not so much interested in the genetics or molecular biology or in the details of x-ray diffraction. We spent some time on the egregiously unfair treatment of the great x-ray spectroscopist Rosalind Franklin, a subject too obvious to be avoided. Most of our seminars focused on the role of hypothesis testing and creativity in this one epochal investigation.

Toward the end of Watson's book, he writes of Francis Crick's announcement of the double helix at the Eagle, a historic Cambridge pub. "Thus I felt slightly queasy when at lunch Francis winged into the Eagle to tell everyone within hearing distance that we had found the secret of life." To the usually cocky Watson, the announcement seemed premature. They had gotten the sugar-phosphate backbone, winding upward like a staircase, exactly right. And between the strands lay the base pairs—flat, steplike, hydrogen bonded. You could climb your way toward whatever stars might lie above, your foot touching first an adenine-thymine pair and then moving to guanine-cytosine and so on, so many possibilities for the base pairs, they were nearly endless.

It seemed perfect. Yet Watson was not quite certain. The bonds between the bases had not been forged in their model which stood in the basement of the Cavendish. It is the same model that you see today in all its architectural perfection, in every textbook, standing against the bare walls of the laboratory. But it did not take long for the atoms and bonds between the base pairs to be crafted according to the best spectroscopic data. When they were, the A-T and G-C pairs were just the right size, as though a carpenter had cut the whole thing, had planed and beveled and sanded. Everyone who saw the model said the same thing: It is too beautiful *not* to exist.

In the darkness and heaviness of the wood-paneled Eagle, I could

170

Photo of the sign outside the "Eagle" in Cambridge.

imagine the young Crick bursting through the door with that swagger and brilliance and self-confidence, that grating loudness that so disturbed Watson, announcing amid the oak tables the perfection of the molecule, created God knows where, in what swamp, in what shallow sea, in what salt lake, like Vanda perhaps, where nothing moved in the depths but stillness itself, or on some distant planet or asteroid undreamed of even by Kepler or Galileo or Newton himself.

And from the Eagle you could take the long sweep through history to a time when no one knew of molecules, when the existence of atoms themselves were challenged and the despair of Boltzmann had deepened to suicide on a bright morning in Duino; to a time when the Earth seemed the center of the universe and stars and planets were immutable, as Aristotle had claimed; long before the chemical bond—let alone the hydrogen bond, which was really a thing of beauty and which in a sense held our watery world, seas and human blood, together—had been recognized for its power to hold and fasten and break. And before a time when anyone knew the power of molecular fluorine, a gas so reactive it could burn hydrocarbons and the coal of my youth without even a spark, and so dangerous the researchers who died working with it were called *fluorine martyrs*.

In the shadows and broken light of the Eagle I recited to myself the last words of the *Double Helix*, the ones about Paris, when Watson is on vacation, the lines that end his book: "But now I was alone, looking at the long-haired girls near St. Germain des Prés and knowing they were not for me. I was twenty-five and too old to be unusual."

Though his hair is white now and Crick is gone, I can only see him as the young man of twenty-five standing in Paris or standing by the sculpture in the Cavendish or following Crick into the Eagle, or the young man composing the first coy lines of their concise nine-hundred-word paper for *Nature*: "We wish to suggest a structure for the salt of deoxyribose nucleic acid (D.N.A.). This structure has novel features which are of considerable biological interest."

❧

Before I caught the train the next day, I went down to the Sedgwick Museum. There, I wanted to think of metals in a different way: not as the tumbling colorless solutes of our Antarctic streams and lakes; I wanted to see them in their trapped vibrancy, locked in the blaze of minerals, sending to the eye some coded beauty that seemed eternal. How much beyond our own would their infinite lives in stone continue?

The Sedgwick, named after the Cambridge geologist Adam Sedgwick who, among much else, was Darwin's professor of geology, though not a supporter of Darwin's theory, is well known for its collection of minerals and for the explanations given for how those minerals formed. It was something in Cambridge I did not want to miss.

In the Whewell Mineral Gallery there were many beryls, which are ringed silicates containing beryllium and aluminum. An emerald is a green beryl in which traces of chromium ion have been substituted for aluminum into the crystal lattice, and red beryl contains traces of manganese. Displayed as they were, these were gorgeous objects of art.

Like the beryls, many of the colors on display at the Sedgwick derived from the presence of trace quantities of the transition metals, the very same metals—chromium, manganese, iron, cobalt, nickel, and copper—that had become our companions in other, less hospitable environments. What distinguished the transition elements from other atoms on Mendeleev's chart and gave them a certain pizzazz were the incomplete d-shells. This made it possible for an element like manganese to absorb light in the blue region of the spectrum and to transmit shades of red.

Our understanding of color in minerals traces back to Bohr and, especially, to the Viennese physicist Erwin Schrödinger, who understood that electrons in atoms occupy orbitals—whispy cloudlike shapes gathered around the nucleus in which the electron has a discrete energy. Electrons in orbitals closer to the positively charged nucleus have lower energies than those farther out. In the case of manganese, the energy difference between two d-orbitals is just right to allow for the absorption of blue light and the transmission of red.

The colors of many other minerals can be explained in terms of transition metal chemistry. So can magnetism, a property which the Cambridge geologists Fred Vine and Drummond Matthews used to test Harry Hess's hypothesis of seafloor spreading, an idea so radical that Hess himself once called it "geopoetry." A lovely poem it turned out to be.

✲

In the small village of Cambridge, not far from London, with its bookstores and bicycles swift as electrons strung along the River Cam, which once seemed so exotic when Clifford spoke its name, history rose from the lawns and sweated from the pores of buildings and pubs that were nearly as ancient as science itself, and all that you had ever learned from *Principia* and *Optics*—the rising of rockets from the Florida cape, hoisting piggyback the gray shuttle, and those failed rockets from your own backyard and childhood dreams, and the teasing of light into strands of color—the visions of Newton lie just over there, in the rooms at Trinity, whose grounds you can point to, its lawns so green—so "fire green as grass," as Dylan Thomas had said—as they touch the Cam. And not far away, the Cavendish, with its prizes and its transformations of the world, the electron, the interior hidden forever deep below the surface, holding the twisted strands, the base pairs of the helix, which one day set the mind to wonderment, to a quest into color and the slowly parting continents of Vine and Matthews and Hess.

I needed to leave, to catch a taxi for the train down to London. And then another train through the tunnel, beneath the sunless Channel, and on to France.

Munich

The first several years we spent in the McMurdo Dry Valleys, we were doing more than just looking at the scant concentrations of trace metals in those distant, pristine waters. At Lake Vanda and the Onyx River, we sank our samplers far into the lake and our cleansed bottles into the river in a search for phosphorus and nitrogen, those elements without which no organism can possibly live. G. Fred Lee had more or less established his reputation on the role that phosphorus plays as a limiting nutrient in lakes and, like myself, he was excited to learn whether or not this same element controlled production in the bizarre waters of Antarctica.

Like many environmental scientists in the early 1960s and '70s, Fred was becoming aware of the changes that were taking place in some of the once pellucid waters around the world. They were becoming infested with massive algal blooms that at the end of their life cycles died and decayed and rotted, and in the process depleted oxygen—often to near-zero levels—from the water. Fish kills, a loss of clarity, and the unforgettable foul stench of rotten eggs from the sulfides were the results. Limnologists called this process *eutrophication*. Eutrophication was on spectacular display in the near-shore waters of Lake Erie, and environmentalists predicted that the lake would soon become a dead zone, incapable of supporting life.

What was causing this? When it came to natural waters, there were hundreds of variables, and without background theory it was difficult to know where to begin. Science as puzzle solving, as Thomas

Kuhn said. But fortunately, in this case, there were clues that suggested that one (or more) of the elements essential to life was being discharged to natural waters in far greater abundance than in the past. It wasn't long before suspicion centered on phosphorus, the backbone element— as Watson and his colleagues had shown at Cambridge—of every DNA molecule on the planet.

It was no secret that human activity was rapidly disturbing the natural phosphorus cycle. Anywhere there was intense agriculture, the phosphorus in fertilizers poured from farm fields into streams and lakes, and to this was added phosphates from detergents and nutrient-laden waters from sewage-treatment plants. It was as though the entire watershed conspired against the lake to shift its phosphorus balance. Limnologists called these human-induced changes "cultural eutrophication," and they were happening rapidly, not at the largo pace of geologic time but in societal time, the paltry years in which we live our lives.

When I taught about eutrophication, I usually mentioned an experiment carried out at the Canada Center for Inland Waters. In one of the many thousands of nameless lakes that dot northern Ontario, scientists divided a single lake in half with a semipermeable membrane. Into one side of the lake they introduced a soluble phosphorus compound. Into the other side, nothing. Then they waited. It wasn't long before the phosphorous-treated lobe began to turn green, lush as a golf course. It was as though a great feast had been laid before the algae, and they glutted themselves as never before. But beyond the barrier, it was life as usual. The only difference between the two sides of the experimental lake was phosphorus—one side rich, the other poor, poor as it must have been for hundreds of years.

Later, in Dallas, G. Fred Lee and his PhD students Walter Rast and Anne Jones were busy collecting all of the data published in journals and reports on phosphorus inputs to lakes. Following a novel and highly regarded model developed by Richard Vollenweider, G. Fred and his young colleagues looked at the connection between phosphorous pouring in from the watershed and a lake's clarity, its algal content, and even the quantity of fish the lake could support. What they learned was that there

Reconstruction of the fission table at the Deutsche Museum in Munich, Germany.

was a clear correlation between phosphorus loading and water quality. Dump in lots of phosphorus and a lake becomes murky. With sparse amounts of phosphorus entering from the watershed, it was just the opposite. The waters were clear and sparkling, as though they had come from a spring. This condition was called *oligotrophy*, and, aesthetically, it was much to be desired. The famous oligotropic lakes, like Superior and Tahoe, were treasured for their beauty, and they lay far to one side of Fred's graph. There were others, like Erie, that lay to the other side. When we completed our study of Acton Lake and placed it on the Vollenweider graph, we found that it was much closer to Lake Erie. And the cause was easy to discern: phosphorus poured in springtime from the surrounding farm fields so that the loading—relative to the area and depth of the lake—was high. My hand became a shadow inches beneath the surface.

To all of this there was a Boltzmannesque theme: nature's power to disperse, to scatter and dismiss, to allow nothing to stay long where you had put it. Fertilizers were made from rocks or from the atmosphere, liberally applied to a field in planting time. And yet it was vain hope to think they would remain there, entering the corn stalks to make them taller. The rains would come and the nutrients would be carried off in a great flood. And the phosphorus and nitrogen would advect into the currents and be swept to the lake and then into other currents in the endless junctures of waterways knitting across the continent and into the sea. And there, in far too many places, in depths shallow and deep, dead zones—lifeless of fish and the sediment-dwelling demersals—were forming in ways that were never intended, never predicted, and, like the bottle of blue gas in the front of the classroom, were forever gone from their containers.

❦

The day I arrived in Munich the rains had stopped, but the River Isar was swollen and nearly overflowing its banks. The rivers were special to me, not so much because I studied them in Antarctica and elsewhere, but because they had a kind of historical permanence to them. After his fame had been well established, the great chemist Justus von Liebig accepted a professorship at the University of Munich, more or less the capital city of southern Germany. He would have known the Isar and its floods and would have probably stood on bridges pondering the roiling water pouring from the snowfields of the Alps.

Liebig is known today for his many contributions to chemistry, but high on the list is his "law of the minimum." This terse statement, of such great importance in ecology, says that a plant will grow in quantity determined by the availability of the element presented to it in least quantity relative to its needs. In lakes in this part of the world, that element was usually phosphorus. It was phosphorus that tended to limit algal production. Not that the soils over which streams flow have little phosphorus in them. Phosphorus is the eleventh most abundant element in the crust. But such is its chemistry that it forms highly insoluble compounds, and these are hardly disturbed by the passage of water. No dissolution. No coaxing into solution. Only a kind of obduracy in the face of water's challenge, its legendary ability to dissolve all that it touches. So in the past, the quantity of algae was kept in check by the grip of phosphorus minerals, which in their miserly way dispensed of their phosphorus atom by atom, never in displays of magnanimity.

Then it changed. Then we learned to cast phosphorus in more soluble forms, to cast it into detergents and fertilizers and waste products, which in time dispersed and wended their way toward lakes and oceans—all of this neatly captured in the nets of number by Vollenweider and then by G. Fred Lee and Walter Rast and Anne Jones in their global survey.

At Lake Vanda, things worked out much as we knew they would. No one had ever occupied the Wright Valley. Phosphorous came from the rocks and soils, so our measurements of the Onyx River were difficult, so small were the concentrations of that element. When we drilled through the lake ice and sank a Secchi disk through the water column, we could see it swaying on the line from side to side like a tiny pendulum. We lost sight of it at twenty-two meters. The water was that clear. We put the phosphorus loading number and the Secchi disk reading on the Vollenweider graph. The point fell high and to the left, higher than Lake Tahoe. If it hadn't been for the four-meter ice cover, we could have seen it at an even deeper depth.

And then we realized we were looking far back at what all lakes must have once been.

❊

In the Hofbrauhaus, I was hungry and warm and back into the rhythms of travel and glad to be inside and taking notes on what I had seen that day. A woman and her friends sat beside me on a long bench, and one of them asked, "Are these your *German reflections*?" and I had a pleasant conversation with her about Munich and the differences between northern and southern Germany. She asked when I was returning to London and I told her I was an American and would be leaving Munich in another day for Bremen and then Paris and the States.

That evening, my "German reflections" included more thoughts about Justus von Liebig and how the law of the minimum applied equally to the land, to the growth of crops, and to the fecundity of the very fields that encircled the towns of West Lafayette and Oxford. The required nutrient that disappeared first from the soils was nitrogen, not phosphorus. We are of course living at the bottom of a sea of nitrogen, which makes up 78 percent of our atmosphere. Nitrogen, which serves as a kind of diluent for the explosively reactive gas, oxygen, is for all practical purposes an inert gas. The diatomic molecule, with its atoms linked together by three strong electron bonds—a triple bond—has been nearly impossible to coax into reaction at normal temperatures. Only the nitrogen-fixing bacteria, the soil's clever chemists, have found a way to do it. Much of German research in the late nineteenth century centered on how to produce synthetic nitrogen fertilizers. It took the long, tedious laboratory studies of Fritz Haber and the engineering skills of Carl Bosch to perfect the reaction between nitrogen and hydrogen required to synthesize ammonia from nitrate so fertilizers could be formed. The geochemist Robert Garrels estimated that the nitrogen taken from the atmosphere by the Haber-Bosch process now equals the nitrogen removed by all of the nitrogen-fixing bacteria on the planet. The long shadow of Justus von Liebig falls across the Earth, from the ten-foot corn in which we lose ourselves in August, as in a maze, to the ice-covered lakes of Antarctica.

There is another, less direct association between Liebig and what we know about the molecular underpinnings of things. Liebig served as mentor to the great organic chemist August Kekulé. We know Kekulé as the man who dreamed the structure of benzene, depending on the version of the story, as a snake catching its tale in its mouth or merely as a whirling presence which became a cycle. Kekulé, whose musings solved the problem of joining six carbons and six hydrogens. Liebig, long before this, saw the brilliance of his young student and convinced him to study chemistry rather than architecture—a subject on which Kekulé had first set his sights. Today we think of Kekulé as the founder of structural chemistry.

The world of molecules, that invisible place below our seeing, is not two-dimensional. It is not simply the composition of a substance written on paper—one of these, four of those, six of something else. Rather, in the wake of Liebig's brilliant student, it is an architecture of forms, a universe of agitation, as Boltzmann had known, a submicron dance of immense complexity, where there is a link between structure and function. The tetrahedral CFC cannot cling to one of its chlorines in the rain of photons that strike it, and so it departs and engages the triangular ozone, and a series of reactions begins. The cocktail of drugs that flow in the veins counter the errant cancer cells, their complex architecture designed to kill.

Liebig lived between 1803 and 1873, and in his lifetime he saw the full efflorescence of organic chemistry and the work of his many students and what he had set in motion, both in Giessen and here in Munich, with its streams swollen and the beer in the Hofbrauhaus flowing as it always had.

※

When she was asked to join the Manhattan Project to build the first atomic bomb, physicist Lise Meitner declined. She wanted nothing to do with bomb making. Yet she was the first to understand what the results that she and Otto Hahn were getting back in the thirties actually meant. The apparatus they used was on display at the Deutsches Museum in Munich. It had been greatly compressed from the original, whose components occupied three rooms. The version I was staring at was on a small table. It consisted of tube and wires and glassware; as humble as a child's toy.

Otto Hahn was a genius with his hands and with his analytical skills. In the apparatus before me, only traces of elements were used, only traces formed. They were nearly counting atoms, one by one, as though they were small invisible gems, bits of diamond in the eye of a scope. But they were infinitely smaller, they gave no hint of their being. The critical experiment involved shooting slow neutrons at a uranium target. There was a Baconian aspect to this, A *let's see what happens if we do x.* As it turned out, strange results occurred. From what was essentially Meitner's apparatus, Hahn and Strassman detected the element barium. This seemed odd, a kind of Kuhnian anomaly based on what had been expected. The expectation was that larger elements, that is, elements with greater atomic mass, transuranics, would be produced. But barium has a relative atomic weight of only 137, compared with 235 for the lighter of the uranium isotopes. Hahn was shocked by the presence of barium in the products. These findings were made in Berlin in 1938, by which time Meitner, who was a Jew, had fled to the safety of Stockholm.

Although Hahn and Strassman reported their results in a paper to *Naturwissenschaften*, they were unable to explain them. It was not until 1939 that Meitner and Otto Frisch communicated their interpretation in a letter to *Nature*. They were the first to realize that a fission reaction had taken place, that the nucleus had been split. It was not long before the significance of this reaction became known: the release of vast stores of energy; the production of neutrons, which would sustain a chain reaction; the potential for a bomb. It was Albert Einstein, in a letter to Roosevelt, who made these frightening prospects known.

I had never heard of Meitner as an undergraduate, never heard of her in Indiana or Virginia. Even the display at the museum until recently referred to the "work table of Otto Hahn." There was no mention of Lise Meitner, its principal designer. In the Deutsche Museum this has finally been changed. Meitner is now included as an equal partner.

There is an element at the very end of the periodic table, where few tend to venture, below iridium, which comes to us from space, the death-dust of dinosaurs, the beginning of our own humble ascendancy, in that column of the table in which reside rhodium and cobalt. The symbol for element 109: Mt. It is meitnerium.

※

The walls of the bar in Bayerstrasse are filled with boxing memorabilia: a robe from Ali's training camp; gloves from George Foreman; an announcement of the Lewis–Schmeling fight, Yankee Stadium, June 22, 1938; the Ali–Frazier fight, Tuesday, September 30, 1975. It is all mixed together: Munich, Schmeling, Hitler, quantum physics, Einstein, Meitner, the bomb. There is an old Coke ad with a girl in a white sweater and a pink bow in her hair, the red of the Coke dispenser, just as I remember it being when, as kids, we were trying to figure how the earth moved. Over in the corner, an old Wurlitzer, its air bubbles moving in arcs along the lemony plastic, like the place in Pittsburgh where we went to dance. Pictures of Monroe, Bogart, the young Dustin Hoffman. A cigar-store Indian stands just inside the door. There is an old-fashioned telephone booth. It is red, with the word TELEPHONE painted on it.

I smelled baked chicken, fries, cigar smoke, and the odd draft of spring air from the streets of Munich. A thin girl in jeans and glasses drank a Coke; her boyfriend in denim drank a beer. They were smoking Marlboros. Do I remember the blackouts over Pittsburgh during the war, or only stories about them? One night Jack Green, my dad's brother, a guy who looked like Brando and died young, pointed out the warships plying the Ohio. He said they had been made at Dravo. It was night and I could just see their lights moving on the dark water. A few years later, I collected war cards. Pictures of Mark Clark, Patton, MacArthur, and my favorite, the Finnish resistance fighters on skis, all in white and carrying riffles. In my parents' dining room, there was the dreamy music of the fifties, before rock and roll: "Harbor Lights," "Goodnight Irene," and anything by Rosemary Clooney.

The kids hated Kuhn. Why? He was right; in science, the revolutions are invisible. There were no parades for the double helix, no flags for Meitner, no upheavals or barricades for Einstein when he reimagined light, or for Rutherford and Bohr when they hollowed out the world. For Copernicus, Kepler, and Galileo, did the crowds cheer and cheer without surcease, did they strike up a band for Edwin Hubble as he came down from Mount Wilson, or for Lavoisier when he told us, finally, how things burned? In the beginning it all looks ramshackle and crazy, like Drake's well in that field by Oil Creek, or like Boltzmann statistics. For fifty years, Alfred Wegener, who saw that the continents really did move, looked like a nut. Most of all to geologists.

Another Kuhnian truth: Science is inherently conservative. Sometimes you just have to pound and pound, like a prize fighter. Here's the ring, here are some gloves. Go at it.

Maybe Meitner saw into the future, beyond Hiroshima and Nagasaki to Ike at Shippensport throwing the switch to the first nuclear power plant. Maybe she saw how medicine would one day use our understanding of the nucleus, or how geology would. There were more than bombs. We still have no idea where this will take us.

In a journey, a lifetime is lived. I did not want to leave Munich.

Oxford, Ohio

My brother asked me to give a lecture on Camus's *The Fall* for his course on religion and ethics. I never thought to ask him why. This was back in Pittsburgh, before he went off to Harvard for his PhD and before his untimely death.

Jean-Baptiste, the narrator, tells his story in the first person to another patron of the Amsterdam bar named Mexico City. Jean-Baptiste was a highly regarded judge in Paris. He was successful in his practice, giving and compassionate in his daily life. He had great regard for himself. But then something happened. One dark night, while crossing the Seine, he heard something strike the water. He knew someone had jumped. In seconds, the screams began. But he continued to walk across the bridge. He never looked down. The screams diminished. Then silence.

Once a lover of heights, of Alpine peaks, of tall buildings, this judge-penitent, as he refers to himself now, has exiled himself to the canal-circled city of Amsterdam, two meters below sea level.

One random event, never dreamed of in youth, had changed everything.

I was not thinking in these terms when I spoke to Eugene's class. Although I mentioned it—the year, his youth—I was not thinking of *randomness* when I told them of Camus's death. He was forty-six. At the last minute, he had decided to travel by car rather by train. At the accident site they found the train tickets stuffed in his pocket.

185

※

In the winter, there was a large snowfall. It came all at once out of a shaking of gray sky.

One morning I took the usual trail past the Dewitt cabin. It was still and there was no one out but me. I wore boots and wind pants like the ones issued at McMurdo. The horses were gone from the pasture, and beyond the fences there was a flat whiteness. The cabin looked sturdy against the night's storm.

Near the trail's end, water poured over the spillway. But on the trees that had been brought downstream in the power of last spring's storms, there was ice on the dead branches and snow piled on the banks.

I had walked an hour and a half when I headed back toward the car. There was a mother and daughter on overland skis. We greeted each other and they moved silently on. It was easier to walk in their tracks than to bury my feet with each step. So the pastures went by quickly, and the cabin with the lives that had been lived there, and I was soon at the car. In five minutes I was home.

As always, there was a peace that came with snow, and I felt transported to another world: Europa, one of the four moons of Jupiter sketched by Galileo in his *Starry Messenger*, a moon of water-ice, with surface patches of ice and snow like an Antarctic lake. When the cancer and the drugs overpowered me so that I could not move, I lay by our picture window and saw the snow in the winter trees. For a while nothing mattered but what lay beyond the glass.

In Pittsburgh, when we had the great snow, I walked the streets at night. There was no one else, no cars. The whole city seemed unreal in its whiteness—a place I had never seen. The silence was broken only by the traffic lights turning red and green.

※

Sometimes, just watching the way that coal burned, how, through the agency of fire, its blackness disappeared in the red glow, the ash vis-

"Kepler" a public sculpture on the grounds of the
Miami University Art Museum in Oxford, Ohio.

ibly forming in the warmth I held my hands toward. Outside, you could hear the sounds of a wooden cart clattering along the sidewalk and the words of the man, "Rags, old iron, rags," became fainter as he moved down the street. Why had the coal changed, its perfect blackness now something else, its hardness come to ash, and the flow of heat upward through the registers, where my aunt sat in her wheelchair?

What was it made of, this lump of coal, shining like glass? And where had it come from? Fern and willow and swamp, the dead sulfiidic waters, the long burial of death through time that we could now measure, through the long efforts of the Curies, Rutherford, Geiger, and so many others.

Decades later, when Barbara and I taught our courses, I read a quote from Einstein, in a speech he gave in honor of Max Planck. It explained why it is that some might want to do science: "A finely tempered nature," he said, "longs to escape from personal life into the

187

world of objective perception and thought; this desire may be compared to the townsman's irresistible longing to escape from his noisy, cramped surroundings into the high mountains, where the eye ranges freely through the still, pure air and fondly traces out the restful contours apparently built for eternity."

✿

My travels taught me that Kuhn had been right about specific revolutions: Einstein, Bohr, the pioneers of atomic structure at the Cavendish. But he was also right about the historical episode that we now call *the* scientific revolution. You wonder how we had ever missed it, how we had been so collectively blind to have not seen what was happening. How the world was being recast in the molds of time, while all around wars raged and madness held our gaze, from which, as though mesmerized, we could not turn away. It cast no shadows, gave not a candle's worth of light, except to those who could hold a mere tube to the night sky and were dumbstruck by its beauty. But who saw beyond the aesthetics—which were enough in themselves—to something else.

Herbert Butterfield said this of the scientific revolution:

Since that revolution overturned the authority in science not only of the middle ages but of the ancient world—since it ended not only in eclipse of scholastic philosophy but in the destruction of Aristotelian physics—it outshines everything since the rise of Christianity and reduces the Renaissance and Reformation to the rank of mere episodes, mere internal displacements, within the system of medieval Christendom. Since it changed the character of men's habitual mental operations even in the conduct of the non-material sciences, while transforming the whole diagram of the physical universe and the very texture of human life itself, it looms so large as the origin of the modern world and of the modern mentality that our customary periodization of human history has become an anachronism and an encumbrance.

Herbert Butterfield died in 1979, before that evening in early January, when high in the thin Martian atmosphere a white flag, a hemisphere of synthetic silk, unfolded against the near horizon. Beneath it, suspended on long cords, a capsule oscillated above the rust-colored surface of the planet. The surface looked much like the landscapes of the McMurdo Dry Valleys.

In time, the oscillations ceased and the whole assemblage rushed toward the hard ground below. After a long parabolic skip-dance across the plains of Mars, the capsule, made of plastic spheres, opened and the rover emerged.

The rover, a robotic Lewis and Clark lighting out for some distant territory, was an emissary of the years since Copernicus, the nearly five hundred years since modern science began with *De Revolutionibus*.

The Mars landing was only one event. But what must we know about the natural world to have done this? What must we know about force and propulsion, about the relationship, in space, of our own Earth to the Red Planet? What must we know about materials, their molecular structure, their strength under enormous strains and insults? About Mars itself, its bedrock composition, its thin unwelcoming atmosphere—mostly carbon dioxide, not nitrogen and oxygen like our own? What truths about the world, in its vastness and smallness, must we have at our fingertips to send an object ten million miles into space and to have it come to rest exactly where we had predicted it would?

I count names, I scramble their history, and I visit the towns and monuments. Only inscriptions in stone, "the stonecutters fighting time with marble," a fight that could not be won. "Of course the stone will never last," Buckley had said, "and Boltzmann knew it."

The five hundred years has been a Cambrian explosion of mind and method. It has allowed us to do what no one, through a million years of trial and error, could ever have conceived possible: walk on the moon; carry on conversations over thousands of miles; build weapons that in a few seconds can obliterate an entire city; design molecules that can alter our blood chemistry; perform calculations at speeds beyond comprehension; fly by the thousands in comfort above cloud and storm into the fading sun; see deep into the human body without surgery, what is there, what fractures, what cells, what imminence of death lurking, that may be countered, held in abeyance. It has allowed us, too, to peer into the universe's beginning, to see its ancient unfolding over time—impossible to grasp—and to read Earth's history, its internal movements, the platforms shifting like a puzzle, the cosmic dangers it has known and will always know.

All of these possible only in the last five hundred years—among protons and electrons and neutrons, the infinite shapes of molecules, the concealment of forces, elusive in torchless caves of darkness, unseen by our crude senses, so often lying anyway, clouding the very motion of the Earth on which we ride.

How long have we been here? Two hundred thousand years. Perhaps. From the time of myth and legend. From the time of dancing torchlight. From elixirs, spells, and magic. It is a fraction of that two hundred thousand years from the Canon of Frauenberg till now. A snap of the fingers from Kepler and Galileo. And with Boltzmann, you feel you could share wine, hang out at Curley's some morning with Rita and Buckley and Harv; or in California, listen to his fingers race across the Steinway.

There was a poll taken recently. Twenty-three percent of us could not name a single scientist; a little more than 40 percent could name only Einstein. There were a few who remembered Carl Sagan. As for Boltzmann, Lavoisier, Arrhenius, they were lost in the obscuring smogs and inversions of the past. For those who were present the day that Katy, Dana, Wanda and I spent at the Griffith in the Hollywood Hills, the percentages would surely have been better, but not by much. Even when you looked down and saw the sprawling city below in all its modernity, you rarely thought to ask where all this came from. What theories, insights, experiments brought this into being? Somewhere, at some juncture, a link with the past had been broken. Perhaps it had never been forged in the first place.

Kuhn's belief that "normal" science was nothing but puzzle solving always disturbed me. It seemed to trivialize science into the Sunday

190

crossword, or into the shuffling of small bits of cardboard that will reveal a mountain.

As I saw it, as long you were adding to human knowledge, as long as your paper spread its wings and landed in the libraries of Paris and Bern and Florence, as long as you were challenging some orthodoxy or someone's treasured beliefs, someone you probably had never met, or as long as your work took you into some fugue state, there among the numbers and figures where you lost track of all time and location, it was so much more than "puzzle solving."

※

Why had Camus not taken the train as he had planned? Why had Ludwig Boltzmann not stayed in Vienna? Why had he not gone to the beach in Duino, instead of choosing to stay by himself? At the Hearst's and in the journals of his travels, he had seemed such a happy man.

Sometimes I see the words of the young Camus as inarguable truths. The ones about those few images from childhood. I look back on my own life, still trying to answer Katy's question posed in the dining hall at McMurdo. Astronomy had carried me away into some new dimension. Not textbook astronomy, but the astronomy of the night sky itself, *felt astronomy*, the astronomy of dreams.

Afterword

One year, after getting back from the Ice, I was asked to coauthor a review paper on the Dry Valley lakes. My friend Berry Lyons and I worked on this for months, trying to condense years of labor and thought into a little less than thirty pages. We began the paper with the words: "Antarctica is the most remote continent on Earth, the most recently discovered, and, hence, the last to feel the human presence." I liked it, in its breadth and somberness, and in all the memories it called to mind.

When I began to write, the front yard, with its many trees, was filled with a foot of snow. It was freshly fallen and white on the leaves, and I could see it through the east-facing window when I glanced up and off to the south in the smaller window near the desk. We began with the lakes of the Taylor Valley, Bonney, and Fryxell, and then moved on to Lake Vanda in the Wright Valley and Joyce in the Pearce Valley. We barely mentioned the work we had done on Lake Hoare and Lake Miers because they were considered freshwater systems, with a relatively low salinity.

I had worked on all of these lakes the year Hatcher, Benoit, and I left from Virginia. But even then, in my barely glancing acquaintance with those diverse and unusual systems, I developed a fondness for Lake Vanda. It was so deep and strange in its chemical structures, and there were mysteries that called from its depths. The waters near the bottom were without oxygen and filled with the poisonous gas hydrogen sulfide. It was like the Black Sea. Had the *Pequod* ever fallen to its bottom in such water, it would have been preserved for thousands of years.

We wrote about the physical and chemical structures of the lakes and about their evolution; we wrote about dissolved organic matter and gas chemistry that Diane McKnight and her colleagues had explored; we wrote about the behavior of nitrogen in Lake Bonney and Fryxell that John Priscu had reported on; we wrote about Berry's predictions on how a warming climate would change the lakes' biogeochemistry; and we wrote about trace metal fate and the chemistry of streams. There was so much more to go. I wanted to do another review on the trace metals and I knew Berry had so much more to say about his own work. We were a long way from that swift survey Benoit, Hatcher and I had conducted.

Our paper swept me so far over time and memory that I was sad to see it in print.

※

Nature loves mixing, dispersal, spread-outness. Always the crushing failure of order: The crude oil pouring from the pipe, unconfined at last, comes to the surfaces roaring, mixes into marshes, encroaches onto beaches, rides the loop currents, encircles the whole peninsula, flows northeastward along the coast of Florida and up toward Cape Cod and then abruptly north toward England. It is the great mixing that, once begun, cannot be stopped, except in the nearly futile skimming and damming with miles of boom, like sausages, over which the tar-balled, wind-driven waves cast their nefarious load. The beaches lie empty, no smell of sunblock; the summer plans, so eagerly made, in ruins. The second law, the law of dispersal, the law on the tombstone in Vienna, seems to guide this unfolding tragedy, a sweet part of earth, the white sand beaches, Antarctic in their purity, given over to greed and ineptitude. Up in Gloucester, far from the demersal torrent and far nearer the calm waters of Thoreau, the fishers know what is coming. The larvae of tuna, on which their lives depend, have depended for generations, are hatched in the now-oily sediments of the Gulf.

In the skies above that same peninsula, there is dispersal in a different way. In a way that is more universal. Oil again, oil that has been refined into one of its many fractions, is being burned in countless engines, sent as water and carbon dioxide and a thousand minor products into the skies. And the prescient Charles David Keeling of Scripps was able to see, as early as the 1950s, that all this burning might add to the atmospheric store of carbon dioxide. His measurements from 1958 onward at Mauna Loa and intermittently at South Pole Station have shown a continuous rise in the concentration of this gas. A gas that intercepts outgoing radiation from the Earth can have only one effect: warming. Had it only been blue, like the fanciful blue gas of my classroom, we might have long ago taken notice.

Mendeleev's visit to Titusville. Drake's primitive well. These small things of no consequence in the Pennsylvania countryside. Who could have known they would mean so much? Boltzmann's body lay in the ground of Vienna. The statistical version of the second law high on the headstone. Who could have seen the connection?

❈

Many things happen in a few years. Not long after Antarctica, Joe died. A dean at the university told me the Antarctic journey had been one of the highlights of his life.

Joseph completed his degree at the University of Hawaii. We coauthored a fine paper on metals in Lake Joyce. He took a position working on a long-term project for the university. He married, and he and his wife purchased a small sailing boat, which they anchored in Honolulu Harbor. They refurbished it and on occasion entered inter-island races.

My daughter Katy received her master's degree in teaching and now works at a private school in Santa Monica. She will be getting married soon to a creative young man who is an animator in the film industry.

I am preparing a "father of the bride" speech, which I am told should be funny and emotional, even to the point of provoking a few tears. If there are tears, I'm hoping they won't be mine.

When Katy and Dana were small, we used to board the train in Chicago and head for Denver. There we rented a car and took it to Breckenridge or Copper Mountain. The first time we did this, Katy was six. It was just after our year in Honolulu. She took easily to the slopes, and after a little instruction was handling the blue runs with ease. By evening she wanted to try the black diamonds, with their steep grades and moguls and leaps and all the crazy fast skiers. I told her it would be better to wait, there was so much we could explore. I told her how I had always loved the fresh snow of early morning high on the slopes.

※

Flying home into Cincinnati from Paris, you can see the big river below. You can see the gentle curve that it makes on its way west to join the Mississippi. And upstream you can see eastward toward the bridge-arched city of Pittsburgh, not the same city of fire and grit in which I grew up. Indiana and its flattened plains lie to the west. Less than two hundred miles away is the city of West Lafayette, where I first heard Walter Edgell tell of a Viennese professor who had committed suicide.

Home in Oxford, I am reminded how time gathers and sifts, gathers and sifts. What remains is small. What burns in memory is smaller still. The rest is scattered, as Boltzmann knew. It will never return. The words of a professor, the rumble of coal down a chute, its sound, its fierce glow in the furnace, its becoming ash. And on starlit nights, the peace of open sky. In snow, the gifts of silence and solitude.

God is geometry, Kepler once said. God is healing, too, I thought. The erratic recovery, the pulling back from the brink, the extra days given as miracle.

A few months after the wedding, I hope to have a beer with Katy. We will talk about those questions she had asked down at McMurdo.

I will tell her about an old quote I had found tacked to a door. I will tell her how the monuments I had seen in my travels had spoken to me of time and history—a history that in some ways was my own. I will tell her about stone and death and the strange valleys I first heard of long ago, valleys that she herself had just seen a few years back. I will remind her of the questions she had asked in the warm galley and how I had never found a clear answer, only an equation inscribed by stone-cutters on a tombstone in Vienna.

Acknowledgments

This book could not have been written without the constant guidance and suggestions of my editor Erika Goldman. She transformed what had been essentially a collection of notes into a coherent and far more personal work. I cannot thank her enough for all the energy and insight that she brought to this project.

I am grateful to Leslie Hodgkins for his careful reading of the late drafts and for the many clarifications that he brought to the book. I also appreciate his excellent cover summary, which I think captures what I hope the reader will find in these pages

My wife, Wanda, read the manuscript in its early stages and along with useful criticisms, provided the moral support that would eventually help bring this to fruition.

Over the years I have received much-needed encouragement in my writing from my sister, Elizabeth Hart, and from many colleagues and friends, including Curt Ellison, Karl Schilling, Xiuwu Liu, Bill Newell and Berry Lyons. It was John Michel who first suggested many years ago that I try writing *Water, Ice and Stone,* and I continue to remain in his debt.

On a recent visit to Los Angeles, my daughter, Dana, introduced me to the Griffith Observatory. I am grateful to her for this introduction, since the Griffith was unexpectedly important in the organization of this book.

This is not written from the perspective of a historian or a physicist (I am neither), but from that of a practicing geochemist with a personal story to tell.

My work in chemistry, geochemistry, and limnology has been guided by my undergraduate inorganic chemistry professor, Bodie Douglas; by Paul E. Field, my wonderful PhD advisor; by the physical chemist Henry S. Frank; and by the aquatic chemist and limnologist G. Fred Lee. I also thank Bob Benoit, Gunter Faure, and Colin Bull for their early roles in my Antarctic research. Professor Berry Lyons has been especially supportive of my work in the McMurdo Dry Valleys. To all of these I owe a great debt of gratitude.

I am extremely grateful to the National Science Foundation for having sponsored my research in Antarctica over the years.

Whatever mistakes the reader my find in this material are solely my own.

❧

I have relied on a number of books that the reader might find enjoyable and instructive. I list a few of these here, in no particular order: Chet Raymo, *The Soul of the Night* and *Honey from Stone*, I. Bernard Cohen, *Birth of a New Physics*; Ruth Lewin Sime, *Lise Meitner*; William H. Brock, *The Norton History of Chemistry*; Banesh Hoffmann, *Albert Einstein: Creator & Rebel*; Charles Tanford and Jacqueline Reynolds, *The Scientific Traveler*; Bertolt Brecht, *Galileo*; Dava Sobel, *The Planets*; John Horgan, *The End of Science*; Kate Farrell, *Art & Nature*; Herbert Butterfield, *The Origins of Modern Science*; James D. Watson, *The Double Helix*; E.J. Dijksterhuis, *The Mechanization of the World Picture*; Arthur Koestler, *The Sleepwalkers*; David Lindley, *Boltzmann's Atom*; Engelbert Broda, *Ludwig Boltzmann*; Elizabeth Kolbert, *Field Notes from a Catastrophe*; C.F. MacIntyre, *Rilke (Selected Poems)*; Bernard Jaffe, *Crucibles: The Story of Chemistry*; Larry Laudan, *Progress and Its Problems*; Thomas Kuhn, *The Structure of Scientific Revolutions*; Sir Isaac Newton, *Principia (Vols. I and II)*; Thomas E. Graedel and Paul J. Crutzen, *Atmosphere, Climate and Change*; Galileo, *Starry Messenger*.

Index